AF248972

Idiotypic Network and Diseases

Idiotypic Network
and Diseases

Editors:

Jan Cerny

Department of Microbiology and Immunology,
University of Maryland School of Medicine,
Baltimore, Maryland

Jacques Hiernaux

GLAXO, Les Vlis, France

American Society for Microbiology
Washington, D.C.

Library of Congress Cataloging-in-Publication Data

Idiotypic network and diseases / editors, Jan Cerny, Jacques Hiernaux.
     p.   cm.
   Includes bibliographical references.
   ISBN 1-55581-025-X
   1. Idiotypic networks.   2. Idiotypic vaccines.   3. Immunotherapy.   I. Cerny, Jan,
1939-   .   II. Hiernaux, Jacques R. J.
   [DNLM: 1. Antibodies, Anti-Idiotypic—immunology.   2. Immunoglobulin Idiotypes—
immunology.   3. Immunoglobulin Idiotypes—physiology.   QW 601 I193]
QR186.3.I35   1990
616.07′9—dc20
DNLM/DLC
for Library of Congress   90-342
                                                                                                                    CIP

CONTENTS

CONTRIBUTORS

M. A. Apicella • Department of Medicine and Department of Microbiology, State University of New York at Buffalo School of Medicine, Buffalo, NY 14215

Jan Cerny • Department of Microbiology and Immunology, University of Maryland School of Medicine, Baltimore, MD 21201

Daniel G. Colley • Veterans Administration Medical Center and Departments of Microbiology and Immunology, and Medicine, Vanderbilt University School of Medicine, Nashville, TN 37212

Nebojsa Dovezenski • Division of Dermatology, Department of Medicine, University of California, San Diego, 225 Dickinson Street, San Diego, CA 92103

Glen N. Gaulton • Department of Pathology and Laboratory Medicine, University of Pennsylvania School of Medicine, Philadelphia, PA 19104

Dorothee Herlyn • The Wistar Institute, 36th Street at Spruce, Philadelphia, PA 19104

Jacques Hiernaux • GLAXO, 25 Avenue du Quebeck, Les Vlis, 11951 Cedex, France

J. Ivanyi • Medical Research Council Tuberculosis and Related Infections Unit, Royal Postgraduate Medical School, Hammersmith Hospital, London W12 OHS, United Kingdom

Joshy Jacob • Department of Microbiology and Immunology, University of Maryland School of Medicine, Baltimore, MD 21201

Garnett Kelsoe • Department of Microbiology and Immunology, University of Maryland School of Medicine, Baltimore, MD 21201

Hilary Koprowski • The Wistar Institute, 36th Street at Spruce, Philadelphia, PA 19104

Petar Lenert • Division of Dermatology, Department of Medicine, University of California, San Diego, 225 Dickinson Street, San Diego, CA 92103

E. Muller • Department of Microbiology, State University of New York at Buffalo School of Medicine, Buffalo, NY 14215

Jean-Marie R. Saint-Remy • Experimental Medicine Unit, Institute of Cellular and Molecular Pathology, Université Catholique de Louvain, 1200 Brussels, Belgium

Maurizio Sollazzo • Division of Dermatology, Department of Medicine, University of California, San Diego, 225 Dickinson Street, San Diego, CA 92103

Anthony J. Weido • Department of Microbiology, University of Texas Medical Branch, Galveston, TX 77550

David B. Weiner • Department of Pathology and Laboratory Medicine and Department of Medicine, University of Pennsylvania School of Medicine, Philadelphia, PA 19104

M. A. J. Westerink • Department of Medicine, State University of New York at Buffalo School of Medicine, Buffalo, NY 14215

Martine Wettendorff • The Wistar Institute, 36th Street at Spruce, Philadelphia, PA 19104

Maurizio Zanetti • Division of Dermatology, Department of Medicine, University of California, San Diego, 225 Dickinson Street, San Diego, CA 92103

PREFACE

Recent advances in molecular techniques have led to the identification and functional characterization of cells and molecules of the immune system. Better tools and approaches are available for the management of immune responses and related diseases. Monoclonal antibodies, recombinant lymphokines, and lymphocyte cloning are just a few examples. However, a rational and predictable immunointervention requires an understanding of the complex interactions between the numerous components of the immune system. A unified concept of immune regulation is still missing. The need for such a concept may help to explain the continuing interest in the idiotypic network hypothesis.

Idiotypes are peptide determinants that are expressed on the variable domains of specific cellular receptors. These include the lymphocyte receptors for antigens as well as the variety of receptors for hormones, viruses, etc., on other cells. The idiotypic network hypothesis is based on the idea that self idiotypes are recognized by the lymphocytes and their specific products. This recognition would provide a unified mechanism for self-regulation of the immune responses and perhaps for interaction between the immune system and other organs. The network hypothesis remains contro versial even though it has been frequently modified to accommodate new immunological knowledge.

This volume opens with a brief discussion of the evolution and complexity of antibody idiotypes. In the next chapter, the editors review the current concepts of the idiotypic network and point out various difficulties in reconciling the hypothesis with present immunological paradigms. In fact, the basic premise of the idiotypic network hypothesis remains rather doubtful. However, experimentation with idiotypic and anti-idiotypic molecules and cells has unraveled new means for manipulating immune responses. The intent of this book is to present a state-of-the-art review of those manipulations and their use in viral, bacterial, and parasitic infections. The expression of particular idiotypes has been used as a disease marker as well as a therapeutic target in autoimmunity, allergy, and neoplastic disease. The new data generated from animal experiments and pre-clinical human studies raise a promise for the use of idiotypic manipulation in medicine.

Chapter 1

Ontogeny of the Antibody Repertoire

Anthony J. Weido, Joshy Jacob, and Garnett Kelsoe

The collection of antibody paratopes constituting the antibody repertoire may be defined in two ways, by genotype or by phenotype. Clearly, the genetic definition is fundamental. The translocations and fusions of DNA necessary to construct the active immunoglobulin H- and L-chain genes specify the diversity available to the antigen-combining site structures (8, 26). Nonetheless, it is phenotypic diversity that alone determines the capacity of the immune response. In this chapter we review the development of the antibody repertoire in C57BL/6 mice, following in parallel the somatic genetics and paratopic phenotypes of the neonate and adult.

ASSEMBLY OF THE ANTIBODY REPERTOIRE

Three classes of genetic elements, V (variable), D (diversity), and J (joining), participate in genomic rearrangements to form the variable region of the immunoglobulin H- ($V_H DJ_H$) and L-chain ($V_\kappa J_\kappa$, $V_\lambda J_\lambda$) genes (reviewed in reference 1). Subsequent pairing of the H- and L-chain polypeptides creates the antibody paratope, or antigen-combining site. Thus, any paratope may be defined as the product of five gene segments, $V_H DJ_H$ and $V_L J_L$.

Germ line V gene segments have been classified into families based upon DNA sequence similarity and occur as reiterated homologs along chromosomes 12 (*Igh*), 5 (*Igκ*), and 16 (*Igλ*) in the mouse. The estimated 150 to 500 V_H gene segments have been classified into 11 homology families in mice (3), while the 100 to 300 V_κ exons have been divided into as many

Anthony J. Weido • Department of Microbiology, University of Texas Medical Branch, Galveston, Texas 77550.   **Joshy Jacob and Garnett Kelsoe** • Department of Microbiology and Immunology, University of Maryland School of Medicine, Baltimore, Maryland 21201.

as 29 families (10). Although significant interspersion of families exists, broadly speaking, families are colocated, defining discrete regions within the *Ig* loci (3, 5). Interestingly, this genomic pattern is not found in all vertebrates but may have arisen from a more primitive genomic arrangement, ... V(D)JC ... V(D)JC ..., found in the primitive shark, *Heterodontus franscisci* (7).

Thus, prior to the introduction of antigen, formation of the primary antibody repertoire may be seen as the product of two related events, rearrangement of the H- and L-chain loci and subsequent association of H and L chains into specific pairs. Processes that control the use of V-region genetic elements or the pairing of the H- and L-chain polypeptides will define the range of genetic variability available for the formation of antibody paratopes. Our laboratory has investigated these processes by determining the frequency at which V_H and V_κ gene families are expressed among mitogen-reactive B lymphocytes taken from the spleens of neonatal or adult mice (10, 11, 15, 21). If mitogen-sensitive B cells represent an unbiased sample with respect to productive recombination of the *Ig* loci and H+L-chain pairing, these studies provide a (partial) somatic genetic description of the primary antibody repertoire.

LYMPHOCYTE CLONING ON FILTER PAPER DISKS

We have used a method developed in our laboratory to investigate the somatic genetics of lymphocyte populations (13). Briefly, suspensions of splenocytes are prepared and plated onto sterile filter paper disks. Subsequently, medium containing thymic feeder cells and a mitogen (generally lipopolysaccharide [LPS]) is added and the cells are cultured for 4 to 6 days at 37°C in a humidified atmosphere of air–5% CO_2. Under such conditions LPS-reactive splenic B cells proliferate to form colonies of daughter cells held immobile within the woven cellulose fibers of the disk. Thus, the method is entirely analogous to the plating of bacteria onto agar medium (Fig. 1). By 5 days of culture, approximately half of all colonies contain at least 100 cells, the number required for reliable in situ hybridization (13). Disks are removed from culture, fixed, dehydrated in ethanol, extracted in chloroform-isoamyl alcohol and air dried (15). They can then be prehybridized overnight at 42°C in a solution containing 50% formamide, 5× SSC (1× SSC is 0.15 M NaCl plus 0.015 M sodium citrate), 5× Denhardt solution, 50 mM phosphate buffer (pH 6.5), 1% glycine, 0.5% sodium dodecyl sulfate (SDS), and ≥50 µg of heat-denatured DNA per ml. Prehybridization solution is then replaced with a hybridization solution containing 50% formamide, 5× SSC, 5× Denhardt solution, 20 mM phosphate buffer (pH 6.5), 0.6% SDS, 7.5% dextran sulfate, and 1×10^5 to 2×10^5

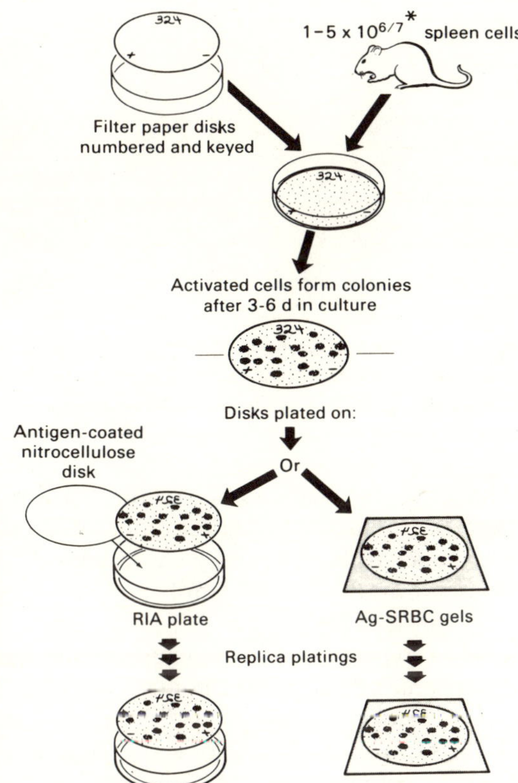

Figure 1. General method for cloning B cells on filter paper disks. Splenocytes (the asterisk indicates maximum densities for LPS/antigen-activated cultures) are plated over thymocyte-impregnated paper disks. After 5 days in culture, adherent colonies of antibody-secreting cells may be identified by various means. Abbreviations: RIA, radioimmunoassay; Ag-SRBC, antigen-sheep erythrocyte (From reference 13.)

cpm of ^{32}P-oligolabeled (Pharmacia, Inc., Piscataway, N.J.) V_H-, V_κ- or C_μ-specific DNA probes per ml (total volume, ca. 4 ml) (10, 15). Routinely, disks are hybridized for 2 days at 42°C and subsequently washed three times in 2× SSC–0.1% SDS at room temperature and once in 0.1× SSC–0.1% SDS at 42°C (15). The disks are then dried and autoradiography is carried out for 5 to 7 days. A stringent wash procedure (20 min in 0.1× SSC–0.1% SDS at 60°C) may be used to strip bound counts for sequential hybridizations.

EXPRESSION OF V_H AND V_κ GENE FAMILIES AMONG LPS-REACTIVE B CELLS

Several groups have investigated the use of V_H and V_κ exons in mice during development (1, 6, 9–11, 21). These studies have revealed that in general, the use frequencies of specific V_H families in adult mice are stoi-

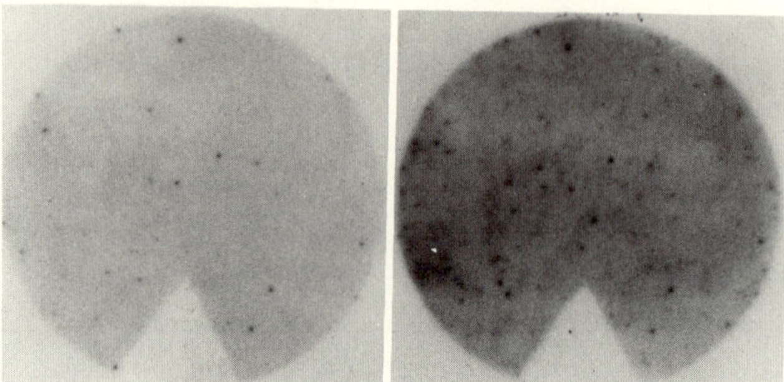

Figure 2. Sequential Northern (RNA) hybridizations to determine the frequency of V_HQ52 expression among $C_\mu{}^+$ colonies. A primary hybridization with a V_HQ52-specific probe (21) identifies $Q52^+$ colonies. Following autoradiography, bound counts were stripped by stringent washing, and the filter was rehybridized with a $C\mu^+$-specific probe. (From reference 13.)

chiometric, or proportional to the V_H family size. For example, Fig. 2 illustrates sequential hybridizations of B-cell colonies grown on a single filter paper disk. Colonies were hybridized first with a probe specific for exons within the V_HQ52 family and subsequently with a probe specific for the μ constant-region gene (C_μ). The primary hybridization identifies only colonies expressing V_HQ52 exons, while hybridization with C_μ identifies virtually all LPS-induced colonies. Thus, the frequency of V_HQ52^+ colonies may be simply calculated as follows: $\%V_HQ52^+$ = (number of V_HQ52^+ colonies/number of C_μ colonies) $\times$ 100. In the example shown in Fig. 2, 23 V_HQ52^+ colonies and 174 $C_\mu{}^+$ colonies were revealed, yielding a use frequency of $(23/174) \times 100\% = 13.2\%$. Based upon the genomic complexity of all known V_H families (2), V_HQ52 exons constitute 11.7% of the entire $Igh\text{-}V$ locus. The concordance between such observed frequencies and the proportional size of V_H families suggests that in the adult mouse virtually any V_H exon is equally likely to participate in productive rearrangements.

Stoichiometric expression is not the rule, however, in the fetal or neonatal mouse (1, 9, 11). Instead, V_H expression is strongly biased for the expression of the 3′ families V_H7183 and V_HQ52. Table 1 summarizes use frequencies for nine V_H families in neonatal and adult C57BL/6 mice with respect to the position of each family (3′ $\rightarrow$ 5′) within the locus. Bias for expression of the V_H7183 and V_HQ52 families in the neonate is readily apparent, as is the normalization of V_H expression in the adult.

In contrast, the expression of V_κ gene segments does not seem to follow the rules of V_H use. V_κ families are expressed at nonstoichiometric frequencies in both neonatal and adult mice (Table 2). Also, whereas neonatal V_κ expression is decidedly biased, no preference for the use of exons located in the 3′ region of the locus is demonstrated. Indeed, if a developmentally

Table 1. Use Frequencies of V_H Gene Families among LPS-Reactive
B Cells from Neonatal and Adult C57BL Mice[a]

V_H family	Complexity[b]	Expression (% of C_μ^+)	
		Neonate	Adult
7183 (3′)	12	25	9
Q52	15	19	13
S107	4	3	3
X-24	2	1	2
36-60	5	6	4
VGAM3-8 (mid)	5	ND[c]	4
J606	10	3	8
3609	16	ND[c]	12
J558 (5′)	62	20	45

[a] From reference 11. Gene order based upon reference 3.
[b] Number of hybridizing restriction fragments observed in a genomic Southern
   blot.
[c] ND, Not determined.

regulated program for V_κ expression exists, it would appear that expression
of exons located in the middle of the locus is favored in the neonate (10).

Tables 1 and 2 illustrate that the expression of both V_H and V_κ exons in
the neonate is strongly biased for only a few gene families. In the neonatal
C57BL/6 mouse, approximately 45% of all C_μ^+ B cells use $V_H 7183$ or
$V_H Q52$ exons and some 75% of B cells express $V_\kappa 1$, $V_\kappa 8$, or $V_\kappa 9$ gene

Table 2. Use Frequencies of V_κ Gene Families among LPS-Reactive
B Cells from Neonatal and Adult C57BL/6 Mice[a]

V_κ family	Complexity[b]	Expression (% of C_μ^+)	
		Neonate	Adult
21 (3′)	10	<1	3
19	10	1	6
10	5	<1	1
8	12	14	5
4 (mid)	8	5	3
9	11	23	5
1	3	40	25
24 (5′)	2	<1	<1
22[c]	7	<1	4
2[c]	5	5	2

[a] From reference 11. Gene order based upon reference 5.
[b] Number of hybridizing restriction fragments observed in a genomic Southern
   blot.
[c] The $V_\kappa 22$ and $V_\kappa 2$ families are unmapped.

segments. As the mouse matures, use frequencies become better distributed among all V_H and V_κ families. The patchy distribution of V-region element use in the neonate must limit the potential diversity of the early antibody repertoire.

RATES OF V_H AND V_κ PAIRING IN LPS-SENSITIVE B CELLS

Expression of H- and L-chain V-region gene segments represents only half of the combinatorial mechanism for generating the primary antibody repertoire. Pairing of the H- and L-chain polypeptides completes the construction of the paratope; influences on the association of specific V_H-V_L combinations would alter the distribution of paratopic phenotypes within the population of antibodies. To investigate the rules of V_H and V_L pairing, we have analyzed V_H and V_κ expression within single B-cell colonies. If pairing is random, that is, if the recombinational choices at one *Ig* locus do not affect those at another, the frequency of any $V_H + V_\kappa$ pair should be the product of the frequencies of independent V_H and V_κ expression. For example, if V_H family X is expressed among 15% of C_μ^+ B cells and V_κ family Y is expressed in 20%, when H- and L-chain pairing is random the expected frequency of $V_H X^+ V_\kappa Y^+$ cells is $0.15 \times 0.20 \times 100\% = 3\%$. Any statistically significant deviation in the observed frequency of $V_H X^+ V_\kappa Y^+$ cells from this expected value implies that coordinate expression of the X and Y gene families is nonrandom, or biased.

Table 3 summarizes the results of sequential hybridizations of B-cell colonies with family-specific V_H and V_κ probes (11). Briefly, disks bearing B-cell colonies were first hybridized with probes specific for the V_HX-24, V_HS107, or V_HQ52 families and autoradiographed. Subsequently, bound

Table 3. Expected versus Observed Frequencies of $V_H^+ V_\kappa$ Cells among Adult C57BL/6 Splenic B Cells[a]

V_H family	% of C_μ^+	Calculated/observed[b] frequencies of V_κ family (% of $C_\mu^\pm$)	
		$V_\kappa 1$ (25.0%)	$V_\kappa 8$ (14.4%)
X-24	1.2%	0.3%/0.3%	0.2%/0.2%
S107	2.2%	0.6%/0.6%	0.3%/0.3%
Q52	10.3%	2.6%/2.5%	1.5%/1.5%

[a] From reference 11. In these experiments 41,330 C_μ^+ colonies were screened in sequential hybridizations for V_H, V_κ, and C_μ expression. Independent V_H and V_κ frequencies were calculated as described previously (13). Colonies coexpressing particular V_H-V_κ pairings were identified as congruent labeled colonies in sequential autoradiograms (15).
[b] Calculated frequencies are derived as percent $V_H \times$ percent V_κ. Observed frequencies are determined as the number of $V_H^+ V_\kappa^+$ colonies/number of C_μ^+ colonies $\times$ 100%.

counts were stripped away by stringent washing and rehybridized with either a $V_\kappa 1$ or $V_\kappa 8$ family-specific probe. After a second autoradiograph, counts were again stripped and disks were hybridized with a C_μ probe. This process resulted in three autoradiographs identifying the positions of V_H^+, V_κ^+, and C_μ^+ colonies on the same filter paper disk. Colonies coordinately expressing a particular V_H-V_κ pairing are easily identified as congruent colonies in sequential radiographs (11, 15). In this way we screened some 4.1×10^4 C_μ^+ colonies for the independent and coordinate expression of three V_H (V_HX-24, V_HS107, V_HQ52) and two V_κ ($V_\kappa 1$ and $V_\kappa 8$) families. Note (Table 3) that at most, the difference between the expected and observed frequencies was only 0.1%. Thus, for this small sample of V_H and V_κ gene families, V_H and V_κ pairing is consistent with a process of random association.

Should random association prove true for all V_H and V_κ exon pairs, we may calculate the expected frequency of any V_H+V_κ pairing, given that the independent V_H and V_κ frequencies are known. This has been done (frequencies from Tables 1 and 2) for 84 V_H+V_κ pairings in neonatal C57BL/6 mice and 108 V_H+V_κ pairings in adult C57BL/6 mice, and the resulting frequencies have been depicted as contour maps (Fig. 3). These maps offer a somatic genetic description of the antibody repertoire and reveal an important characteristic of the distribution of V_H+V_κ pairs among neonatal and adult mice. Of the 84 pairings presented for neonates, five gene families (V_H7183, V_HQ52, V_HJ558, $V_\kappa 1$, and $V_\kappa 9$) account for nearly 40% of all LPS-reactive splenic B cells. The distribution of V_H+V_κ pairings is more disperse in adult mice: only 2 of the 108 pairs studied represent frequencies higher than 4%. Thus, neonatal C57BL/6 mice utilize, on average, fewer V_H+V_κ pairings to construct antibody paratopes than do their adult counterparts. This developmentally regulated pattern of V_H and V_κ expression may be responsible for the ordered acquisition of humoral immune responsiveness observed in mice (22) and many other species (18, 20, 23).

MAPPING PARATOPIC SPECIFICITY TO V-REGION GENE SEGMENTS

Particular V_H or V_L gene segments have been associated with specific immune response. For example, phosphorylcholine (PC) in immunogenic form regularly elicits an antibody response in mice that nearly exclusively uses a single member of the V_HS107 gene family, V1 (4). This dominance and the remarkable phylogenetic preservation of the V_HS107 gene family (17) have led to the suggestion that these genes have evolved specifically for optimal reactivity to PC (19).

To extend this logic, if evolution has modified V-gene families for

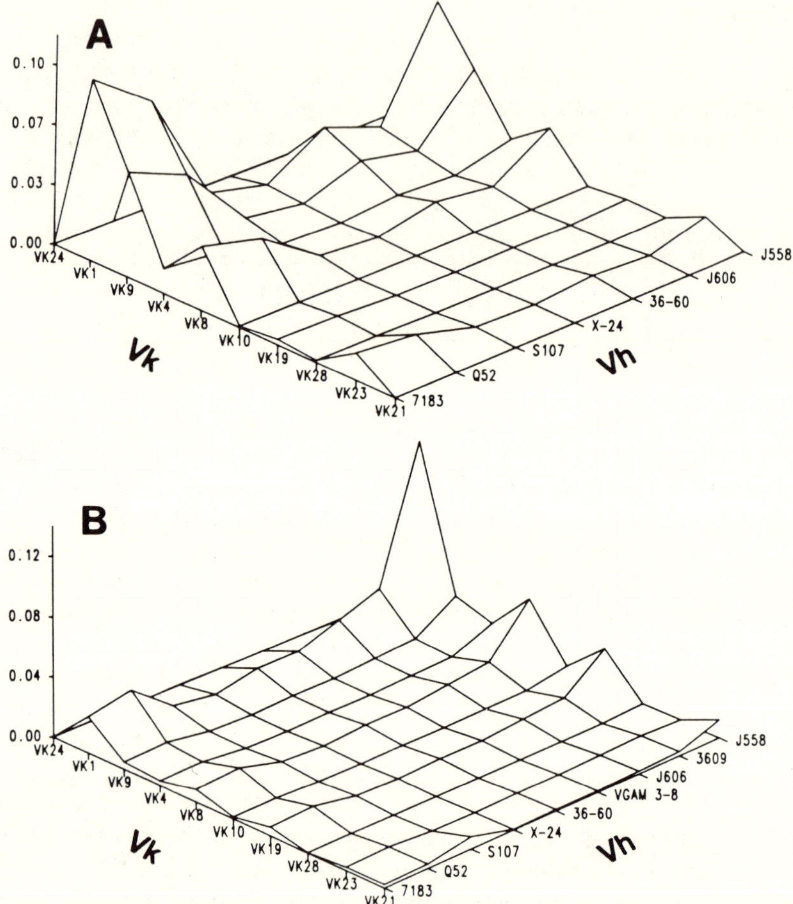

Figure 3. Contour maps depicting the relative frequencies of V_H-V_κ pairing in neonatal (A) and adult (B) C57BL/6 mice. Note the highly skewed distribution of V_H-V_κ pairings in the neonate as opposed to the more nearly uniform distribution in the adult.

specific reactivities, nonrandom associations between V-gene family expression and paratopic specificity should occur even among the primary, naïve, repertoire. We have tested this prediction by determining the antibody specificity and V_H gene family expressed by LPS-induced colonies of murine B cells (15). Briefly, filter paper disks bearing colonies of splenic B lymphocytes were blotted onto nitrocellulose sheets coated with specific antigens, PC, or hen egg lysozyme. Colonies that secreted antibody specific for (a single epitope of) the bound antigen left a footprint of bound immu-

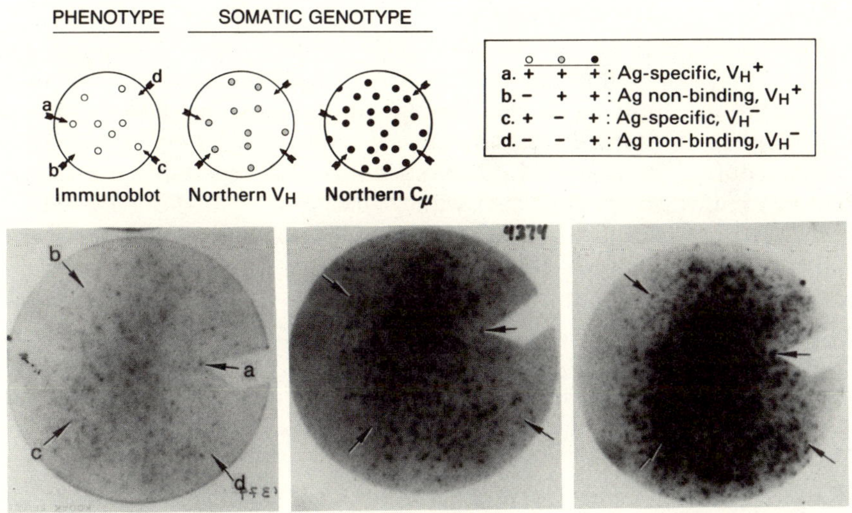

Figure 4. Method for determining V_H to paratope associations by sequential phenotypic and genotypic analyses. An example of the procedure is illustrated below the schematic. The illustrated sequence was PC; V_HJ558; C_μ. (From reference 15.)

noglobulin. Subsequently, disks were fixed and V_H and C_μ expression was determined as described above. Footprints of specifically bound immunoglobulin were visualized by incubation of washed nitrocellulose blots with ^{125}I-labeled goat anti-mouse κ antibody (13). Finally, autoradiographs of the serial immunoblot and hybridizations may be compared to identify V_H expression among antigen-specific colonies (Fig. 4).

The results of these experiments are summarized in Table 4. Note that among unselected, C_μ^+ colonies, expression of the V_HX-24, V_HS107,

Table 4. Failure To Map Antibody Specificity to V_H Gene Families in the Primary Antibody Repertoire[a]

Colony type	% of $C_\mu^\pm$ colonies expressing:			
	V_HX-24	V_HS107	V_HQ52	V_HJ558
Unselected C_μ^+	1.0	3.0	8.9	36.9
PC-specific C_μ^+	1.0	1.0	9.65	35.4
HEL-specific $C_\mu^{+\,b}$	1.7	1.7	10.1	42.0
id$^+$/PC-specific C_μ^+	ND^c	17.9	ND^c	12.0

[a] From reference 15.
[b] HEL, Hen egg lysozyme.
[c] ND, Not determined.

V_HQ52, and V_HJ558 gene families is stoichiometric (see also Table 1). Surprisingly, stoichiometric expression also occurs among antigen-specific colonies. These data imply that in the primary antibody repertoire of adult C57BL/6 mice, similar paratopic specificities may be generated by all V_H gene families, presumably in different (V_H) DJ$_H$ + V_LJ$_L$ contexts. As a positive control, PC-specific C_μ^+ colonies that also expressed an idiotope (B36-82) known to be associated with PC-binding antibodies encoded by the V_HS107 gene family (24) were analyzed. In this case, expression of the dual (PC/id$^+$) phenotype clearly mapped positively to V_HS107 expression and negatively to V_HJ558 expression (Table 4). This demonstrates that failure to map individual V_H families to particular antibody specificities was not due to insensitivity of the mapping procedure. We must conclude that all V_H exons have the capacity to encode PC- or hen egg lysozyme-specific paratopes. Presumably, this ability depends on the great diversity of (V_H) DJ$_H$ + V_LJ$_L$ combinations available. Indeed, it may be calculated that, given the random association of V-region genetic elements (see above), each V_H exon may participate in some $1V_H \times 10D \times 4J_H \times 250V_\kappa \times 4J_\kappa = 4 \times 10^4$ unique κ^+ antibodies. Intuitively, it is not surprising that, given this degree of potential diversity, each V_H family is competent to create paratopes specific for many distinct antigenic determinants. Diversity may be the character that evolution has brought to the primary repertoire instead of specificity. The optimal primary repertoire may be one of high cross-reactivity and relatively low specificity (25). Indeed, even in mouse strains bearing different *Igh-V* loci, the repertoire and frequencies of paratopic specificities are virtually identical (16).

WHAT IS THE MEANING OF DEVELOPMENTALLY REGULATED CHANGES IN THE ANTIBODY REPERTOIRE?

Our studies have demonstrated an orderly process of age-associated V_H and V_κ gene family expression in the developing C57BL/6 mouse (Tables 1 and 2). The distribution of $V_H + V_\kappa$ pairings in the neonate is significantly different from that in the adult (Fig. 3). This program of gene expression is correlated with a well-described patterned acquisition of humoral immune responsiveness (22). It seems likely that the two events represent cause and effect. The antibody repertoire of the neonatal mouse lacks the profound diversity found in the adult animal.

Does the acquisition of paratopic specificity (and increasing antibody diversity) necessarily imply an immunological significance to the process? Or is the process merely the epiphenomenon of the mechanics of opening the *Ig-V* loci to transcriptional activity? For example, Kearney et al. (12) have argued that the fetal and neonatal repertoire is biased for the expression of anti-idiotypic reactivities and that the resulting early interactions between lymphocytes represent an important step in shaping the antibody repertoire. Although this may indeed be true, several conceptual difficulties arise. All suggestions that the ontogenically early repertoire is biased in some way imply that certain paratopic specificities (e.g., anti-idiotypes) map to specific clusters of V_H and V_L genes. We have been unable to demonstrate such associations for at least four V_H gene families, including one, V_HQ52, that is overexpressed in the fetus and neonate (Table 4). Secondly, even if early interactions between mutually reactive lymphocytes do affect the developing antibody repertoire, they do so in such a way that V_H expression and the association of H and L chains in the adult are apparently random (Tables 1 and 3). Finally, the paratopic phenotype that results from these processes also appears to be random: paratopic frequencies for protein antigens are highly correlated with the surface area of the antigen ligand rather than being measures of foreignness (14, 25, 27; A. J. Weido and G. Kelsoe, unpublished observations) or even association with a pathogen (J. E. Ubelaker and G. Kelsoe, unpublished observations).

We take these findings to suggest that evolution has not acted to ensure the expression of any particular set of paratopic reactivities. Although the neonatal antibody repertoire contains much less diversity than does its adult counterpart, we believe that this most probably represents an epiphenomenon of chromosome structure rather than an evolutionarily conserved programming of expressed antibody specificities.

REFERENCES

1. **Alt, F., T. K. Blackwell, and G. D. Yancopoulos.** 1987. Development of the primary antibody repertoire. *Science* **238:**1079–1087.
2. **Brodeur, P. H.** 1987. Genes encoding the immunoglobulin variable regions, p. 81–109. *In* F. Calabi and M. S. Neuberger (ed.), *Molecular Genetics of Immunoglobulin.* Elsevier/North-Holland Publishing Co., Amsterdam.
3. **Brodeur, P. H., G. E. Osman, J. J. Mackle, and T. M. Lalor.** 1988. The organization of the mouse Igh-V locus. Dispersion, interspersion, and the evolution of V_H gene family clusters. *J. Exp. Med.* **168:**2261–2278.
4. **Clarke, S. H., J. L. Claflin, M. Potter, and A. Rudikoff.** 1983. Polymorphisms in antiphosphorylcholine antibodies reflecting evolution of immunoglobulin families. *J. Exp. Med.* **157:**98–113.
5. **D'Hoostelaere, L. A., K. Huppi, B. Mock, C. Mallet, and M. Potter.** 1988. The immunoglobulin kappa light chain allelic groups among the Igκ haplotypes and Igκ crossover populations suggest a gene order. *J. Immunol.* **141:**652–661.
6. **Dildrop, R., U. Krawinkel, E. Winter, and K. Rajewsky.** 1985. V_H-gene expression in murine lipopolysaccharide blasts distributes over the nine known V_H-gene groups and may be random. *Eur. J. Immunol.* **15:**1154–1156.
7. **Hinds, K. R., and G. W. Litman.** 1986. Major reorganization of immunoglobulin V_H segmental elements during vertebrate evolution. *Nature* (London) **320:**546–549.
8. **Honjo, T.** 1983. Immunoglobulin genes. *Annu. Rev. Immunol.* **1:**499–528.

9. **Jeong, H. D., and J. M. Teale.** 1988. Comparison of the fetal and adult functional B cell repertoires by analysis of V_H gene family expression. *J. Exp. Med.* **168**:589–603.

10. **Kaushik, A., D. H. Schulze, C. Bona, and G. Kelsoe.** 1989. Murine V_κ gene expression does not follow the V_H paradigm. *J. Exp. Med.* **169**:1859–1864.

11. **Kaushik, A., D. H. Schulze, F. A. Bonilla, C. Bona, and G. Kelsoe.** 1989. Stochastic pairing of $V_H + V_\kappa$ occurs in polyclonally activated B cells. *Proc. Natl. Acad. Sci. USA,* in press.

12. **Kearney, J. F., V. M. Vakil, and D. S. Dwyer.** 1987. Idiotypes and autoimmunity. *CIBA Found. Symp.* **129**:109–123.

13. **Kelsoe, G.** 1987. Cloning of mitogen- and antigen-reactive B lymphocytes on filter paper discs: phenotypic and genotypic analysis of B-cell colonies. *Methods Enzymol.* **150**:287–304.

14. **Kelsoe, G., and D. Farina.** 1987. Is the antibody repertoire biased? p. 163–174. *In* G. Kelsoe and D. H. Schulze (ed.), *Evolution and Vertebrate Immunity.* University of Texas Press, Austin.

15. **Kelsoe, G., R. Miceli, J. Cerny, and D. H. Schulze.** 1989. Mapping of antibody specificities to V_H gene families. *Immunogenetics* **29**:288–296.

16. **Kelsoe, G., and J. T. Stout.** 1986. Cloning of mitogen- and antigen-reactive B lymphocytes on filter paper discs. II. Paratope frequencies within the mitogen-selected repertoire. *Cell. Immunol.* **98**:506–516.

17. **Litman, G. W., L. Berger, K. Murphy, R. Litman, K. Hinds, and B. W. Erickson.** 1985. Immunoglobulin V_H gene structure and diversity in *Heterodontus,* a phylogenetically primitive shark. *Proc. Natl. Acad. Sci. USA* **82**:2082–2086.

18. **Lydyard, P. M., C. E. Grossi, and M. D. Cooper.** 1976. Ontogeny of B cells in the chicken. I. Sequential development of clonal diversity in the bursa. *J. Exp. Med.* **144**:79–88.

19. **Perlmutter, R. M., B. Berson, J. A. Griffin, and L. E. Hood.** 1985. Diversity in the germline antibody repertoire. Molecular evolution of the T15 V_H gene family. *J. Exp. Med.* **162**:1998–2016.

20. **Rowlands, D. T., D. Blakeslee, and E. Angala.** 1974. Acquired immunity in opossum (*Didelphis virginiana*) embryos. *J. Immunol.* **112**:2148–2159.

21. **Schulze, D. H., and G. Kelsoe.** 1987. Genotypic analysis of B cell colonies by *in situ* hybridization: stoichiometric expression of three V_H families in adult C57BL/6 and BALB/c mice. *J. Exp. Med.* **166**:163–172.

22. **Sherwin, W. K., and D. T. Rowlands.** 1975. Determinants of the hierarchy of humoral immune responsiveness during ontogeny. *J. Immunol.* **115**:1549–1557.

23. **Silverstein, A. M., J. W. Uhr, K. L. Kramer, and R. J. Lukes.** 1963. Fetal response to antigenic stimulus. II. Antibody production by the fetal lamb. *J. Exp. Med.* **117**:799–808.

24. **Strickland, F. M., J. T. Gleason, and J. Cerny.** 1987. Serologic and molecular characterization of the T15 idiotype. II. Structural basis of independent idiotope expression on phosphorylcholine-specific monoclonal antibodies. *Mol. Immunol.* **24**:637–646.

25. **Striebich, C. C., R. M. Miceli, D. H. Schulze, G. Kelsoe, and J. Cerny.** 1990. Antigen-binding repertoire and immunoglobulin heavy chain gene usage among B cell hybridomas from normal and autoimmune mice. *J. Immunol.,* in press.

26. **Tonegawa. S.** 1983. Somatic generation of antibody diversity. *Nature* (London) **302**:575–581.

27. **White-Scharf, M. E., M. Souroujon, J. Andre-Schwartz, R. S. Schwartz, and M. L. Gefter.** 1988. Specificity of the germline-encoded pre-immune B cell repertoire, p. 105–117. *In* O. N. Witte, N. R. Klinman, and M. C. Howard (ed.), *B Cell Development.* Alan R. Liss, Inc., New York.

Chapter 2

Concept of Idiotypic Network: Description and Functions

Jan Cerny and Jacques Hiernaux

The immune system is composed of a large number of functionally distinct cells, receptors, and effector molecules. Multiple interactions take place within the immune system, which is functionally linked to other organs. The immune response must be well regulated because minor alterations of the immunoregulatory pathways may lead to various pathological situations. Therefore, a precise understanding of the immunoregulatory processes and of the nature of the pathological alterations could lead to the development of immunotherapeutic manipulations as well as to better means of immunoprophylaxis. However, a unified conceptual framework that would integrate the bulk of experimental data does not exist, and the immunoregulatory process is not fully understood. The idiotypic-network hypothesis, formulated by N. Jerne more than 15 years ago (31), aimed to provide a simple set of rules for self regulation of the immune response. The hypothesis was based on the dual character of the antibody molecule: it recognizes an antigen through its combining site (the paratope), and it is "immunogenic" by virtue of its idiotopes. Idiotopes are antigenic structures (epitopes) present on the variable (V) portion of the antibody molecules of all the mammals studied so far. Each antibody molecule carries a set of multiple idiotopes that defines its unique idiotype. It has been established that an animal can be immunized with purified, autologous antibody molecules to produce specific anti-idiotope, suggesting that the self-idiotope determinants are indeed recognized by the immune system. Jerne's hypothesis postulated that vast repertoires of antigen epitopes and idiotopes structur-

Jan Cerny • Department of Microbiology and Immunology, University of Maryland School of Medicine, Baltimore, Maryland 21201.   **Jacques Hiernaux** • GLAXO, 25 Avenue du Quebeck, Les Vlis, 11951 Cedex, France.

ally overlap, such that a given antibody molecule reacts with an antigen as well as with an idiotope of another antibody molecule. Thus, the antibody-producing lymphocytes form a network in which the clones of cells expressing distinct immunological specificity interact with and regulate each other's function.

The role of the idiotypic network in immune regulation remains controversial; however, Jerne's hypothesis has stimulated a large amount of experimental research. Much has been learned about (i) the immune responses against the immunoglobulin idiotypes, (ii) the characterization of idiotypes and idiotopes by serological means, (iii) the molecular structure of idiotope determinants, and (iv) the manipulation of the immune responses with specific idiotypic and anti-idiotypic reagents. These topics have been reviewed in several articles and monographs (5, 6, 15, 46). The present volume deals with the recent research on the role of idiotypic interactions in microbial infections and in various immunologically related diseases. Nonetheless, each chapter covers some basic aspects of idiotypy. The purpose of the general introductory section is to provide the background information for the reader. We shall briefly discuss the idiotype network hypothesis and the structure and function of idiotypes in the remaining parts of this chapter.

THE IDIOTYPIC NETWORK

In his original presentation, Jerne (31) viewed the idiotypic network as a B-cell network in which interlymphocytic connections depend on complementary V-region structures present on antibody molecules. In essence, an idiotype is recognized by the antibody combining site, or paratope, of an anti-idiotype. Figure 1 presents Jerne's original view of the network. An antigenic epitope (E) induces an antibody, Ab_1, recognizing the epitope through its paratope, p_1, and expressing an idiotype, Id_1 (i_1 in Fig. 1). Id_1

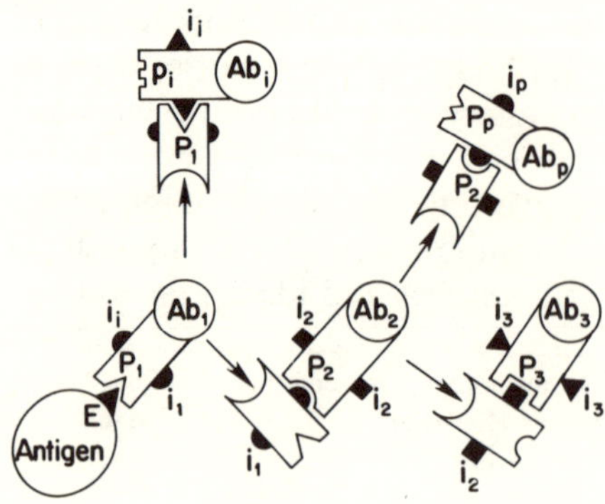

Figure 1. Jerne's idiotypic network (31). A foreign epitope, E (▲), triggers an immune response, Ab_1. The paratope, p_1, recognizes the epitope, and the idiotype, i_1, induces an anti-idiotypic response, Ab_2. p_2 recognizes i_1, and i_2 stimulates an anti-anti-idiotypic response, Ab_3; p_3 recognizes i_2. Ab_i, the internal image antibody, expresses an idiotype, i_i, recognized by p_1. Ab_p, the parallel-set antibody, expresses i_1 ($i_p = i_1$), recognized by p_2, and equates a distinct antigenic specificity through its paratope, p_p.

induces Ab_2, which possesses Id_2 and recognizes the Id_1 through p_2. Id_2 induces Ab_3, and so on. . . . Jerne also introduced two other elements: (i) the internal image antibody, Ab_i, expressing the idiotype Id_i and mimicking the original antigenic epitope (E); the paratope of Ab_i recognizes an unidentified antigen; (ii) the parallel-set antibody, Ab_p, expressing the idiotype Id_1 and possessing the paratope p_p, which recognizes an unknown epitope (different from E). The Ab_i and Ab_p antibody molecules have been identified experimentally.

One of the major problems with the original presentation of the network is that T lymphocytes, which are key regulatory elements of the immune system, are not explicitly involved. A first attempt to incorporate T cells into the network was made by Hoffmann (28), who developed a theoretical model including T cells secreting monovalent suppressor T factors. Later, Hiernaux and Bona (25), as well as Urbain and co-workers (35, 60), presented other regulatory schemas involving T lymphocytes. Those models were developed before the more recent data became available concerning the structure of the T-cell receptor (TCR) and the mode of antigen presentation. The TCR is a heterodimer composed of two chains (α and β or γ and δ) presenting a V-region domain. Idiotypelike structures have been identified on those V regions. In terms of the network connectivity, it seems rather unlikely that B cells and T cells that respond to the same antigen would have common idiotypes, because B cells and T cells usually recognize different epitopes. Indeed, B-cell receptors (i.e., the immunoglobulin molecules) recognize native epitopic structures, whereas TCRs recognize processed epitopes associated with the class I or class II major histocompatibility complex (MHC) molecules. Earlier experiments with polyclonal anti-idiotypic antibody had suggested a sharing of idiotypic specificity between B cells and T cells, but those initial observations have rarely been confirmed in more recent work with monoclonal antibody. On the other hand, one can conceive the existence of anti-idiotypic B cells and T cells that recognize different epitopes of the same idiotype. One can thus envision a B-cell network with the participation of anti-idiotypic T lymphocytes at each level. The latter could be either a helper T (CD4) or cytotoxic T (CD8) lymphocyte. We have chosen not to discuss the role of suppressor T lymphocytes in this chapter, since they are not as well defined as the other two categories of T lymphocytes. In Fig. 2 we present a network scheme compatible with the most recent immunological data. Indeed, idiotype-specific T-cell clones have been described, and complementary idiotypic and anti-idiotypic antibody molecules have been isolated as monoclonal antibodies in syngeneic animals. Nevertheless, the existence of anti-idiotypic B and T lymphocytes does not imply that the idiotypic network is physiological. Elements relevant to that question are discussed in Autologous Regulation of Immune Response by Idiotypic Mechanisms. The scheme

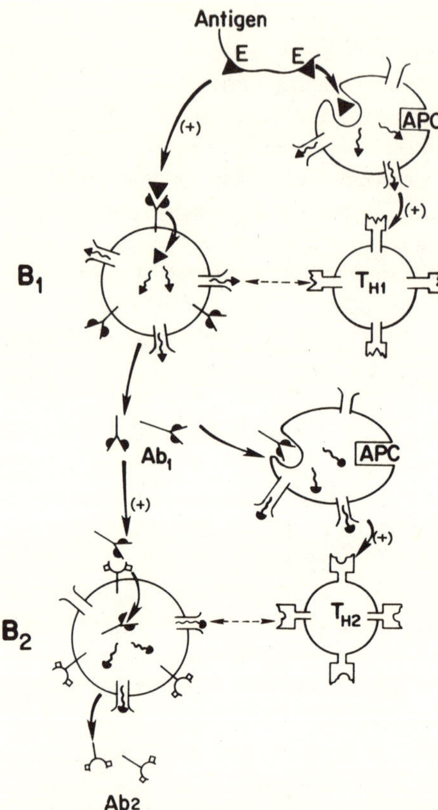

Figure 2. An antigenic epitope (▲) primes B₁ lymphocytes expressing Ab₁ surface immunoglobulin (sIg). An antigen-presenting cell (APC) processes the antigen and presents an epitopic peptide ( ) in association with an MHC class II antigen ( ) ( ). This complex ( ) induces the proliferation of T_H₁ cells, which recognize the same complex on the surface of B₁ cells. This allows the activation of B₁ cells by T_H₁ cells. Ab₁ antibody ( ) primes B₂ lymphocytes, expressing Ab₂ sIg ( ). The processing of an Ab₁ molecule by an antigen-presenting cell generates idiotopic peptides ( ), which can associate with MHC class II antigens. The complex ( ) can trigger T_H₂ cells, which are able to recognize the complex on B₂ lymphocytes. This recognition leads to further activation of B₂ cells and production of Ab₂. This schema does not include the possibility (64) that the antigen-reactive B₁ cells process their own sIg and express the MHC-idiotope complex ( ). In such case, the B₁ lymphocyte could also receive help from the idiotope-reactive T_H₂ cell.

of Fig. 2 is still traditional in the sense that we adopted Jerne's original concepts (especially the dual character of antigen-recognizing structures) to more recent immunological data. It is, however, important to mention that recent experimental studies indicate that the distinction between a paratope and an idiotype might be purely functional and that, depending on the experimental design, an idiotype might behave like an anti-idiotype and vice versa. The degeneracy of the paratope-idiotype duality leads to the "sticky-end" concept introduced by Jerne in 1982 (32). Köhler et al. have recently suggested (40) that alternative idiotypic-network models should be developed on the basis of this concept.

THE NATURE OF IDIOTYPES

The Structure of Immunoglobulin Idiotopes

As discussed by Weido et al. (this volume), the V region of the immunoglobulin molecule is assembled from five protein segments encoded by distinct sets of genes: V_H, D_H, and J_H segments of the heavy (H) chain and V_L and J_L peptides of the light (L) chain. The number of different genetic segments within each set varies from 4 J_H genes to >100 V_H genes, for a total of $\geqslant 10^3$ genes. The germ line-encoded antibody repertoire is generated by random combination of these genes, with one exon from each of five sets. The number of different antibody molecules that can be generated in this manner is >10^6. Additional molecular diversity is produced at the points of joining of different DNA segments.

As a result of the genetic combinatorial mechanism of antibody diversity, two immunoglobulin molecules with different binding specificity (i.e., different paratopes) may share a peptide segment (for example, V_L) encoded by the same gene. This may be the structural basis for the existence of the "parallel-set" antibodies (Ab_p) (see The Idiotypic Network). The reverse is also true: two immunoglobulin molecules with the same paratope may use a different peptide segment.

The structural basis of idiotypy has been the subject of several reviews (13, 21, 37, 46, 48). The primary amino acid sequences of the V segments (including the joining sequences) represent the linear (continuing) idiotopes. Such idiotope determinants are recognized by specific monoclonal antibodies (anti-idiotope) on the whole immunoglobulin molecule as well as on isolated H and L chains and on synthetic peptides corresponding to the V segments. The linear idiotope determinants may be found on immunoglobulin molecules that differ in the paratope but share a particular V segment. Moreover, one can envision that these may be processed and recognized by the T lymphocytes as the peptides associated with the MHC molecules (see Gaulton and Weiner, Colley, and Westerink et al., this volume).

The second class of germ line-encoded idiotopes comprises the conformational determinants that are produced by the tertiary structures of the V-region domains and by the quaternary interactions between the H and L chains. The monoclonal anti-idiotope specific for the conformational determinants binds to the intact immunoglobulin but not to the isolated H and L chains. An assignment of the conformational idiotope to specific amino acid sequences is difficult; indeed, the expression of these idiotopes may be influenced by amino acid substitutions in different segments of the protein chain. The possible recognition of such complex epitopes by the T lymphocytes is difficult to envision. It may be that certain T cells recognize the native immunoglobulin molecules and their conformational idiotopes; however, this notion is inconsistent with the current paradigm. Alternatively, the conformational idiotope may be processed and presented to the lymphocytes in a manner that maintains their structure, which would place significant constraints on the mechanisms of protein processing. The evidence suggesting that T cells may recognize the conformational immuno-globulin idiotope will be discussed by Gaulton and Weiner, Colley, and Westerink et al. (this volume).

The rearranged immunoglobulin genes undergo somatic mutations in the course of the antigen-driven immune response; this leads to the emergence of antibody molecules with altered paratopes as well as idiotopes. The B-cell clones expressing immunoglobulin with the highest affinity for the antigen are preferentially stimulated and expanded to dominate the antibody response. This process of affinity maturation of immunoglobulin molecules often leads to an idiotypic switch whereby newly emerging idiotypic determinants may replace the early, germ line-encoded idiotype. These non-germ line idiotypes may be either linear or conformational, depending on the site(s) of the amino acid replacement(s). It remains to be established whether the somatic antibody variants may be recognized and selected by an autologous anti-idiotype response within the immune network.

Designation of Idiotypic Determinants

The discrete epitopes on the V region are usually called idiotopes; each idiotope is identified by a specific monoclonal antibody. The collection of individual idiotopes forms the idiotype of the immunoglobulin molecule. Prior to the development of the monoclonal antibody technology, an idiotype was defined by conventional, polyclonal antisera that reacted against several distinct idiotopes on the immunoglobulin molecule. The abbreviations Id and id have been used in the literature for both idiotope and idiotype. The conventional nomenclature of these determinants, which is used in this volume, is summarized in Table 1.

It has been estimated that there are 15 to 20 idiotopes on the V region of a single immunoglobulin molecule (45), which may be distinguished by corresponding monoclonal anti-idiotope (i.e., the Ab_2, in the parlance of the network hypothesis). Some of these determinants are topographically distant from the antigen-binding site (paratope); they have been dubbed nonparatopic or framework idiotopes, and the antibodies that recognize them are sometimes called $Ab_2\alpha$. The binding of $Ab_2\alpha$ to the idiotope is not inhibited by the antigen. Other epitopes that overlap with the binding site, partially or fully, are called near-paratopic and paratopic idiotopes, respectively, and the corresponding anti-idiotopes have been designated as $Ab_2\gamma$ and $Ab_2\beta$ (7, 33). The reaction of $Ab_2\beta$ with the idiotope is antigen inhibitable, suggesting that the business end of $Ab_2\beta$ mimics the antigen itself; i.e., it is an internal image of the antigenic epitope (see Fig. 2 of Wettendorff et al., this volume).

The localization of idiotope on a given antibody molecule may have some implications for the design of the corresponding anti-idiotope vaccine.

Table 1. Conventional Nomenclature of Idiotypic Determinants

Name of idiotype (synonyms)	Description	Abbreviation(s)	Corresponding anti-idiotype
Cross-reactive (public, recurrent)	Expressed on antibodies from most individuals within a strain or species (it can be either a major or minor idiotype)	CRI, IdX	
Major (dominant, regulatory)	Found on a large fraction of serum antibody (usually expressed as CRI)	(CRI_M)	
Minor	Found on a small fraction of serum antibody (it can be expressed as CRI)	(CRI_m)	
Private	Found on some individual antibodies	IdI, Idi	
Paratopic	Overlaps with the antigen-binding site (paratope)		$Ab_2\beta$
Near-paratopic	Located close to the antigen-binding site		$Ab_2\gamma$
Nonparatopic (framework)	Located distant from the antigen-binding site		$Ab_2\alpha$
Internal image	Idiotype on the $Ab_2\beta$	$Id\beta$	

There is no evidence to suggest that $Ab_2\beta$ are intrinsically more potent activators of the antibody response than are $Ab_2\alpha$. However, it is reasonable to assume that antibody molecules that have different genetic origins but the same specificity (paratope) would react with $Ab_2\beta$ (that mimics the antigen) more predictably than they would with $Ab_2\alpha$. Thus, $Ab_2\beta$ may have a broader reactivity with the antigen-specific B-cell clones from various individuals. On the other hand, the binding of $Ab_2\alpha$ to the antigen-reactive lymphocytes does not block the paratope of the surface immunoglobulin, which suggests the theoretical possibility that both the anti-idiotypic antibody and the antigen participate in the regulation of antibody formation at the level of B-cell activation. The considerations for choosing various anti-idiotypic antibodies as vaccines are discussed in detail by Wettendorff et al. (this volume).

The number of different idiotopes that appear in response to immunization is determined by the diversity of the response, i.e., the heterogeneity of the germ line-encoded antibody, the process of somatic mutation, and the selection of various lymphocyte clones. These mechanisms are influenced by the nature of the antigen and the genetic makeup of the host. The idiotypic variability of antibody response is reflected in the operational definition of idiotypes and idiotopes according to their representation

(Table 1). Certain idiotopes appear on a large portion of antibody molecules from each individual animal within an inbred strain; such determinants have been called cross-reactive (CR idiotype; CRI), dominant, recurrent, or public. For example, the antibody response to *Streptococcus pneumoniae* R36a is so well conserved that the dominant idiotype, T15, is expressed on a large proportion of antibody molecules from all mouse strains and, perhaps, from other mammalian species (11a). In contrast, some idiotypes appear on a very small fraction of the antibody such that they are detectable only in some individual inbred animals. These idiotypes are referred to as minor or private. It has been speculated that the frequently occurring idiotypes may play a more important role in the immune network regulation, for which the term regulatory idiotype was coined (5).

Idiotypes of T Cells

The unique epitopes (idiotopes) expressed on the TCRs have not been studied as thoroughly as the immunoglobulin idiotopes. It is clear that the V domains of various TCR α/β and γ/δ heterodimers express multiple idiotope determinants that are detectable with the clonotype-specific antisera (i.e., conventional antibody made against the antigen-specific T-cell clones) and, recently, with the monoclonal antibodies that are specific for the peptides encoded by various V genes of the TCR gene complex (42). A number of investigators have explored the role of the T-cell-associated idiotypes in network regulation. It has been reported that the specific suppressor T-cell subsets interact with one another as idiotypic (Id$^+$) and anti-idiotypic lymphocytes possessing the complementary receptors (14, 57); however, this work awaits confirmation. Studies of cell-mediated responses to MHC antigens have demonstrated the role of idiotypic T cell–T cell interaction in regulation of the graft-versus-host reaction and allograft rejection (38, 56, 66). Injection of alloantigen-activated T cells has been shown to suppress the host cellular response to the antigen, which is mediated, presumably, by the activation of anti-idiotypic T cells (1). The structural relationship between the TCRs of certain MHC-reactive T cells and their complementary anti-idiotypic T cells has been studied by Sim et al. (52). Cohen (12) has recently described several T-cell clones that are specific for the basic myelin protein and that produce a degenerative neurological disease upon transfer into normal syngeneic rats. The animals that were previously immunized with these autoreactive T-cell clones developed a specific anti-clonotypic immunity that was mediated by the host T lymphocytes and that rendered them resistant to the disease.

Relationship between the Immunoglobulin and T-Cell Idiotopes

The immunoglobulin and TCR molecules are genetically distinct, but their structural motifs are similar. Indeed, several reports have described an idiotopic similarity (cross-reactivity) between the immunologically related, antigen-specific B cells and T cells (10, 15, 23, 41). For example, Ertl and Finberg (17) injected mice with a monoclonal anti-idiotope against the Sendai virus-specific immunoglobulin and observed that the animals developed the T-cell response against the virus. Similar observations have been made in other systems (reviewed in reference 18), and some of these results are discussed by Gaulton and Weiner and Wettendorff et al. (this volume). However, all studies suggesting the idiotopic sharing between the immunoglobulin and TCR molecules to date are based on the functional effects of anti-idiotopes on the T cells. We do not know of any confirmed report of the immunochemical isolation of TCR with the immunoglobulin-specific, monoclonal anti-idiotope. Moreover, the concept of idiotope cross-reactivity between the immunoglobulin and TCR is made doubtful by the accepted fact that the B cells and T cells recognize different epitopes on a given protein antigen and that the antigen binding of the TCR, but not the immunoglobulin, includes the MHC molecule. Considering the large number of idiotope determinants borne on the vast repertoire of specific antigen receptors, it is not surprising to find that a given anti-clonotypic (TCR) antibody cross-reacts with an immunoglobulin; however, the biological significance of such cross-reactivities remains to be established.

Recognition of Immunoglobulin Idiotypes by T Cells

Immunization of mice with a syngeneic, monoclonal immunoglobulin that bears a defined idiotype leads to the activation of T cells that recognize the idiotype. These anti-idiotypic T cells may provide specific help for (16, 30, 43) or suppression of (51) the antibody response against haptens coupled to the Id^+ immunoglobulin. In other words, the T cells appear to recognize the self idiotype in a manner similar to the recognition of a foreign carrier protein. Indeed, the subsequent analysis of T-cell clones revealed that the idiotype recognition is restricted by the self MHC molecules (4, 49, 53) and that these anti-idiotypic T cells are specific for linear peptides resulting from the processing of the V regions of immunoglobulin H or L chains (22, 34, 51, 64).

Collectively, the published data demonstrate that (i) T cells have the potential to react against self idiotypes; (ii) anti-idiotypic T cells can be activated by the artificial means of immunization with the idiotype molecules, and (iii) the T cells recognize small fragments of the V domains. However, there is no evidence that these T cells become activated physiologically in the course of the antigen-driven immune response. Moreover, the T-cell repertoire does not appear to include the unique, conformational idiotypic determinants that are recognized by antibodies and that are involved primarily in the antibody-mediated network regulation. The apparent dichotomy between anti-idiotype antibody and T-cell repertoires makes it difficult at present to envision a unified idiotypic network of T cells and B cells.

THE PRINCIPLES OF IDIOTYPIC MANIPULATION OF THE IMMUNE RESPONSE

Stimulated by the network hypothesis, a large volume of experimental work has been performed that demonstrated that anti-idiotypic antibody is able to modulate the antibody response expressing the corresponding idiotype. Various factors influence the outcome of the idiotypic manipulation: (i) the nature of the target idiotype (i.e., private or public), (ii) the nature of the anti-idiotypic antibody (monoclonal versus polyclonal, syngeneic, allogeneic, or xenogeneic), (iii) the use of an adjuvant, (iv) the coupling of the anti-idiotypic antibody to an immunogenic carrier (it is also important to realize that xenogeneic Fc can induce a strong carrier effect), (v) the loss of anti-idiotypic antibody, (vi) the time of injection (i.e., prenatal, neonatal, or adult), and (vii) the isotype of the anti-idiotypic antibody. Since there are a large number of variables, it is often difficult to predict the outcome of an idiotypic manipulation. Another unpredictable aspect at present is the nature of the regulatory loops, either suppressive or enhancing, induced by the anti-idiotype. The most obvious effect should result from the interaction between the anti-idiotype and the idiotype-positive (Id^+) surface immunoglobulin (sIg) present on the B lymphocytes: the cross-linking of the sIg induces the regulatory signals turning the target B cell on and off. The experiments of Trenkner and Riblet demonstrated a dose-dependent, bell-shaped curve of B-cell stimulation by anti-idiotypic antibody in vitro (59). The effects of anti-idiotypic antibody on B cells in vivo may be influenced by the regulatory T lymphocytes, in a manner that remains obscure at present. Nevertheless, some manipulations seem to have a predictable outcome. Indeed, the idiotypic cascade concept, initially developed by Urbain et al. (61) and Cazenave (8), has predictable consequences. Immunization with xenogeneic or allogeneic anti-idiotypic antibody (so-called Ab_2) in an adjuvant leads to the production of anti-idiotypic (Ab_3) antibody. A fraction of Ab_3 shares some idiotopic specificities with the original Ab_1. In some cases, Ab_3 also binds the original antigen. Moreover, the Ab_3-producing animals always produce Ab_1-like antibodies after subsequent challenge with the original antigen. The Ab_1-like antibodies share idiotypic and paratopic specificities with the first, antigen-induced antibody. This type of manipulation has allowed the induction of a rabbitlike Ab_1 (expressing epitopic and idiotypic specificities never seen in mice) in Ab_3-producing mice (19).

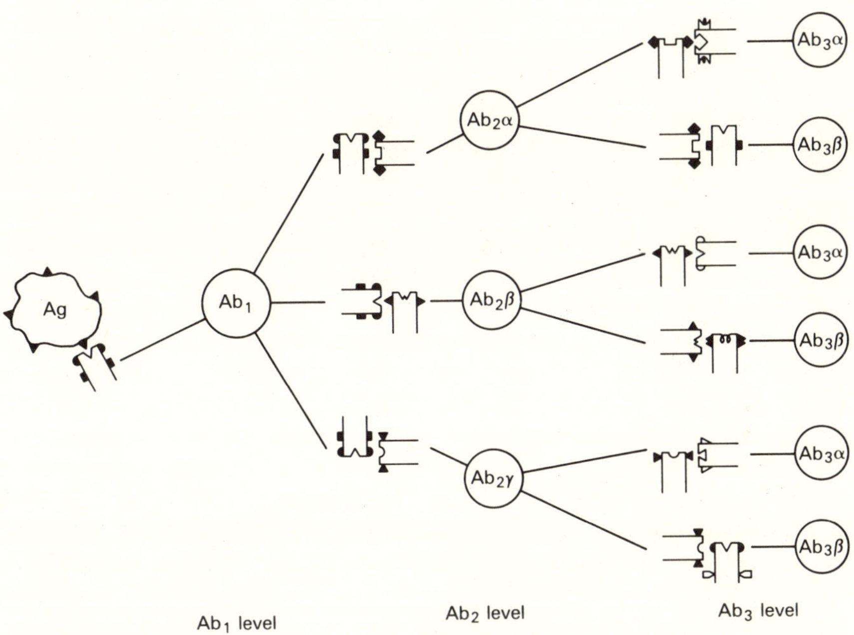

Figure 3. Idiotypic cascade. A foreign epitope (▲) induces an immune response characterized by the production of Ab₁ antibody. Ab₁ elicits an anti-idiotypic response (Ab₂), which contains three subsets: Ab₂α recognizes a framework-associated idiotype (■) on Ab₁; Ab₂γ recognizes an antigen-combining site-related idiotope (▲) on Ab₁; and Ab₂β presents the internal image of the original antigenic epitope. Each subset of Ab₂ can trigger an anti-anti-idiotypic response (Ab₃). The Ab₃ response is fairly complex, and the nature of the Ab₃ depends on the inducing Ab₂. Some of the Ab₃ subsets are shown. Ab₃β presents an internal image of the epitope inducing the corresponding Ab₂; a subset of Ab₃β antibodies induced by Ab₂α or Ab₂γ are Ab₁-like. Ab₃α antibodies are directed against the idiotopes of the various Ab₂ antibodies; Ab₃α antibodies induced by Ab₂β have the same antigen-combining site as Ab₁. For each type of antibody molecule, one V region is shown.

Figure 3 presents some of the elements of the idiotypic cascade (24) as well as their potential relationships. For the sake of simplicity, the presentation is limited to a few key elements. The antigenic epitope induces the production of Ab₁. The Ab₁ expresses a series of idiotopes that can be divided on the basis of their topography. Some idiotopes are closely related to the antigen-combining site, while other idiotopes are associated with the immunoglobulin framework. The Ab₁, in turn, triggers an anti-idiotypic response consisting of distinct types of Ab₂ antibodies (18, 19). Ab₂α recognizes the framework-associated idiotopes, Ab₂γ recognizes the idiotopic determinants closely associated with the paratope, and Ab₂β bears an idiotope mimicking the antigenic epitope (i.e., it is an internal-image

Ab_2). Each subset of Ab_2 can, in turn, induce distinct subsets of anti-anti-idiotypic antibody, Ab_3, and so on. Various factors probably restrict the extension of these network reactions; if this were not the case, the immune system would be degenerate. The relative proportion of the various Ab_2 subsets ($Ab_2\alpha$, $Ab_2\beta$, and $Ab_2\gamma$) depends on unknown variables including, perhaps, the heterogeneity of the Ab_1 molecule(s) produced in response to the antigen.

The concept of anti-idiotypic vaccines has emerged in the 1980s. Initially, it was proposed to use $Ab_2\beta$ as a vaccine because the idiotope of this antibody mimics the antigenic epitope and it should induce primarily an Ab_1-like response (44). Nevertheless, $Ab_2\alpha$ and $Ab_2\gamma$ are also capable of inducing Ab_1-like antibody and/or priming the recipient for response to the antigen (46, 50). However, the immunization with $Ab_2\alpha$ may also induce an antibody that shares a framework-associated idiotope with Ab_1 but does not bind the antigen.

The concept of idiotypic cascade has been derived from experiments with adult animals having the mature immune system. Administration of either Ab_1 or Ab_2 during early ontogeny may have different, profound effects on the development of antibody responses. The antibody repertoire may be divided into two parts: (i) the expressed repertoire, which is composed of naturally occurring antibodies and antibodies (Ab_1) that are normally produced in response to antigenic stimulation; and (ii) the "silent" repertoire, which consists of lymphocyte clones that have the receptors for antigen but do not engage in antibody production under normal circumstances. Both parts of the antibody repertoire may be influenced by the idiotypic manipulation of fetuses and neonates. The immature immune system is characterized by a high degree of idiotype–anti-idiotype interactions and is more sensitive to the idiotypic manipulations than the adult immune system is (36). For example, a neonatal injection of Ab_2 may result in a lifelong suppression of antibody (Ab_1) response (55), or it may prime the neonate for a higher response to the antigen challenge later in life (54). On the other hand, neonatal treatment with a low dose of Ab_2 may lead to activation of the silent Ab_1-producing clones (26).

Idiotypic manipulation of pregnant females influences the antibody repertoire of the offspring (65), suggesting that the immune responses during pregnancy may have an active role in the development of the child's immune system.

In conclusion, it is fair to say that the effects of idiotypic manipulation (i.e., injection of Ab_1 or Ab_2) are unpredictable and that the treatment remains empirical. A better understanding of the idiotypic interactions, immune regulation, and ontogeny of idiotypes is required for a rational design of idiotypic vaccines.

AUTOLOGOUS REGULATION OF IMMUNE RESPONSE BY IDIOTYPIC MECHANISMS

There is no doubt that the immune system contains the lymphocytes that recognize and respond to the idiotypes of the self immunoglobulin molecules. The evidence lies in the fact that immunization of animals with either syngeneic or autologous idiotypes produces the idiotype-specific antibody as well as the idiotype-reactive T cells. It has been also shown that the experimentally induced anti-idiotype antibody and T cells have the ability to regulate the functions of the idiotype-bearing (Id$^+$), antigen-reactive lymphocytes. Thus, it would appear that the key components of the idiotype network hypothesis are in place: the antigen stimulates the Id$^+$ lymphocytes to proliferate and to produce the Id$^+$ molecules, which, in turn, activate the autologous anti-idiotype lymphocytes that regulate the antigen-driven, Id$^+$ immune response. However, the experimental evidence for such an autologous physiological network regulation remains unconvincing, despite years of investigation. It may be that the putative autologous anti-idiotype responses are so weak that their measurements are unreliable and subject to technical artifacts, unlike the strong responses induced by experimental immunization with the idiotype.

The presence of anti-idiotype (Ab$_2$) in the sera of antigen-stimulated animals and humans has been reported by a number of investigators (reviewed in references 20, 39, and 47). These anti-anti-idiotype molecules, however, have not been properly purified and characterized. Moreover, it has been very difficult to identify the cells producing the autologous Ab$_2$; Hiernaux et al. found very few plaque-forming cells producing Ab$_2$ in mice immunized with bacterial levan (27), and they were concerned about the specificity of the assay. The most troublesome fact is that, to our knowledge, nobody has been able to obtain any Ab$_2$ hybridomas by fusion of lymphocytes from antigen-stimulated animals. Nothing short of isolation of autologous anti-idiotypic monoclonal antibodies can provide convincing support for the network hypothesis. Interestingly, the hybridomas producing various autologous anti-idiotypic antibodies have been easily obtained from the lymphoid tissues of fetal and neonatal mice (36). This finding leads some investigators to believe that there are autologous anti-idiotypic antibody responses during the early developmental stages that may influence the immune repertoires of the individual and that cease to exist in the mature organism.

The activity of autologous anti-idiotypic T cells in the course of immune response also remains doubtful. There have been several reports suggesting that the bulk T cells, isolated either from normal donors or at some time after antigen administration, possess the specific anti-idiotypic activity that can be demonstrated by the ability of the T cells to regulate the antigen-specific response of the Id$^+$ lymphocytes (2, 7a, 9, 29, 58). However, these putative idiotype-specific T cells have never been established as lines and clones and characterized. Idiotype-specific T cell lines and clones have been recently obtained from a concanavalin A-stimulated lymphocyte library (3) and from animals that had been actively immunized either with purified Id$^+$ immunoglobulins (4) or with syngeneic Id$^+$ B cells (11, 62, 67). These clones should become valuable in dissecting the repertoire of idiotype-reactive T cells. It appears that some of the T cells (which have been activated by the Id$^+$ B cells) may recognize the intact, unprocessed immunoglobulin molecule rather than the small peptides (11). The activation of such lymphocytes by an autologous idiotype recognition, however, remains a hypothesis.

The most impressive, albeit indirect, evidence for autologous idiotypic regulation has been obtained in the course of chronic parasitic infections and is discussed by Colley (this volume). It is tempting to speculate that the idiotype-reactive lymphocyte clones remain relatively silent during a normal, transient immune response but become activated under the extreme conditions of ongoing immunological stimulation by chronic infection. Such well-characterized disease models may eventually provide the ultimate test of the network hypothesis.

REFERENCES

1. **Anderson, L. C., H. Binz, and H. Wigzell.** 1976. Specific unresponsiveness to transplantation antigens induced by antoimmunization with syngeneic, antigen-specific T lymphoblasts. *Nature* (London) **264:**778–780.

2. **Becker-Dunn, E., and K. Bottomly.** 1985. T15-specific helper T cells: analysis of idiotype specificity by competitive inhibition assay. *Eur. J. Immunol.* **15:**728–732.

3. **Becker-Dunn, E., J. Kim, and K. Bottomly.** 1986. A cloned T cell line that selectively augments antibody responses of phosphorylcholine-specific B cells bearing the T15 idiotype. *J. Mol. Cell. Immunol.* **2:**209–217.

4. **Bogen, B., B. Malissen, and W. Haas.** 1986. Idiotype-specific T cell clones that recognize syngeneic immunoglobulin fragments in the context of class II molecules. *Eur. J. Immunol.* **16:**1373–1378.

5. **Bona, C.** 1987. *Regulatory Idiotopes.* John Wiley & Sons, Inc., New York.

6. **Bona, C., and J. Hiernaux.** 1981. Immune response: idiotype antiidiotype network. *Crit. Rev. Immunol.* **2:**33–81.

7. **Bona, C., and H. Köhler.** 1984. Anti-idiotypic antibodies and internal images, *Recept. Biochem. Methodol.* **14:**141–149.

7a. **Bona, C., and W. E. Paul.** 1979. Cellular basis of regulation of expression of idiotype. I. T-suppressor cells specific for MOPC 460 idiotype regulate the expression of cells secreting anti-TNP antibodies bearing 460 idiotype. *J. Exp. Med.* **149:**592–600.

8. **Cazenave, P. A.** 1977. Idiotypic-anti-idiotypic regulation of antibody synthesis in rabbits. *Proc. Natl. Acad. Sci. USA* **74:**5122–5125.

9. **Cerny, J., and M. J. Caulfield.** 1981. Stimulation of specific antibody-forming cells in

antigen-primed nude mice by the adoptive transfer of syngeneic anti-idiotypic T cells. *J. Immunol.* **126**:2262–2266.

10. **Cerny, J., C. Heusser, R. Wallich, G. J. Hammerling, and D. D. Eardley.** 1982. Immunoglobulin idiotypes expressed by T cells. *J. Exp. Med.* **156**:719–730.

11. **Cerny, J., J. S. Smith, C. Web, and P. W. Tucker.** 1988. Properties of anti-idiotypic T cell lines propagated with syngeneic B lymphocytes. I. T cells bind intact idiotypes and discriminate between the somatic idiotypic variants in a manner similar to the anti-idiotopic antibodies. *J. Immunol.* **141**:3718–3725.

11a.**Claflin, J. L., J. Wolfe, A. Maddalena, and S. Hudak.** 1984. The murine antibody response to phosphocholine. Idiotypes, structures and binding site, p. 171–195. *In* M. I. Greene and A. Nisonoff (ed.), *The Biology of Idiotypes.* Plenum Publishing Corp., New York.

12. **Cohen, I. R.** 1986. Regulation of autoimmune disease: physiological and therapeutic. *Immunol. Rev.* **94**:5–22.

13. **Davie, J., M. V. Sieden, N. S. Greenspan, C. T. Lutz, T. L. Bartholow, and B. L. Clevinger.** 1986. Structural correlates of idiotopes. *Annu. Rev. Immunol.* **4**:147–165.

14. **Dorf, M. E., K. Okuda, and M. Minami.** 1982. Dissection of a suppressor cell cascade. *Curr. Top. Microbiol. Immunol.* **100**:61–67.

15. **Eichmann, K.** 1978. Expression and function of idiotypes on lymphocytes. *Adv. Immunol.* **26**:195–224.

16. **Eichmann, K., I. Falk, and K. Rajewsky.** 1978. Recognition of idiotypes in lymphocyte interactions. II. Antigen-independent cooperation between T and B lymphocytes that possess similar and complementary idiotypes. *Eur. J. Immunol.* **8**:853–857.

17. **Ertl, H. C. J., and R. W. Finberg.** 1984. Sendai virus-specific T cell clones: induction of cytolytic T cells by an anti-idiotypic antibody directed against a helper T cell clone. *Proc. Natl. Acad. Sci. USA* **81**:2850–2854.

18. **Finberg, R. W., and H. C. J. Ertl.** 1986. Use of T cell-specific anti-idiotypes to immunize against viral infections. *Immunol. Rev.* **90**:129–155.

19. **Fraucotte, M., and J. Urbain.** 1984. Induction of anti-tobacco mosaic virus antibodies in mice by rabbit anti-idiotypic antibodies. *J. Exp. Med.* **160**:1485–1494.

20. **Geha, R. S.** 1986. Idiotypic interactions in the treatment of human disease. *Adv. Immunol.* **39**:255–297.

21. **Greenspan, N. S., and W. J. Monafo.** 1987. Topographic analysis with monoclonal anti-idiotopes: probing the functional anatomy of immunoglobulin variable domains. *Int. Rev. Immunol.* **2**:391–417.

22. **Hannestad, K., G. Kristoffersen, and J. P. Briand.** 1986. The T lymphocyte response to syngeneic 2 light chain idiotopes. Significance of individual amino acids revealed by variant 2 chains and idiotope-mimicking chemically synthesized peptides. *Eur. J. Immunol.* **16**:889–893.

23. **Harvey, M. A., K. Adorini, A. Miller, and E. E. Sercarz.** 1979. Lysozyme-induced T-suppressor cells and antibodies have a predominant idiotype. *Nature* (London) **281**:594–596.

24. **Hiernaux, J.** 1988. Idiotypic vaccines and infectious diseases. *Infect. Immun.* **56**:1407–1413.

25. **Hiernaux, J., and C. Bona.** 1980. Immune network, p. 269–297. *In* C. Bona and P. A. Cazenave (ed.), *Lymphocytic Regulation by Antibodies.* John Wiley & Sons, Inc., New York.

26. **Hiernaux, J., C. Bona, and P. J. Baker.** 1981. Neonatal treatment with low doses of antiidiotypic antibody leads to the expression of a silent clone. *J. Exp. Med.* **153**:1004–1008.

27. **Hiernaux, J. R., J. Chiang, P. J. Baker, C. Delisi, and B. Prescott.** 1982. Lack of involve-

ment of auto-anti-idiotypic antibody in the regulation of oscillations and tolerance in the antibody response to levan. *Cell. Immunol.* **67**:334–345.

28. **Hoffmann, G. W.** 1975. A theory of regulation and self-nonsex discrimination in an immune network. *Eur. J. Immunol.* **5**:638–647.

29. **Janeway, C.** 1986. Varieties of idiotype-specific helper T cells: a commentary. *J. Mol. Cell. Immunol.* **2**:265–267.

30. **Janeway, C. A., Jr., N. Sakato, and H. N. Eisen.** 1975. Recognition of immunoglobulin idiotypes by thymus-derived lymphocytes. *Proc. Natl. Acad. Sci. USA* **72**:2357–2360.

31. **Jerne, N. K.** 1974. Towards a network theory of the immune system. *Ann. Inst. Pasteur Immunol. Sect. C* **125**:373–389.

32. **Jerne, N. K.** 1982. *Idiotypes—Antigens on the Inside,* p. 238–239. Editiones 'Roche', Basel.

33. **Jerne, N. K., J. Roland, and P. A. Cazenave.** 1982. Recurrent idiotopes, and internal images. *EMBO J.* **1**:243–247.

34. **Jorgensen, T., and K. Hannestad.** 1982. Helper T cell recognition of the variable domains of a mouse myeloma protein (315). Effect of the major histocompatibility complex and domain conformation. *J. Exp. Med.* **155**:1587–1596.

35. **Kaufman, M., J. Urbain, and R. Thomas.** 1985. Towards a logical analysis of the immune response. *J. Theor. Biol.* **114**:527–561.

36. **Kearney, J. F., and M. Vakil.** 1986. Idiotype-directed interactions during autogeny play a major role in the establishment of the adult B cell repertoire. *Immunol. Rev.* **94**:39–50.

37. **Kieber-Emmons, T., and H. Köhler.** 1986. Towards a unified theory of immunoglobulin structure-function relations. *Immunol. Rev.* **90**:29–48.

38. **Kimura, H., and D. B. Wilson.** 1982. Anti-idiotypic cytotoxic T cells in rats with graft-versus-host disease. *Nature* (London) **308**:463–464.

39. **Köhler, H.** 1980. Idiotypic network interactions. *Immunol. Today* **1**:18–21.

40. **Köhler, H., T. Kieber-Emmons, S. Srinivasan, S. Kaveri, W. J. Morrow, S. Muller, C. Y. Kang, and S. Raychaudhuri.** 1989. Revised immune network concepts. *Clin. Immunol. Immunopathol.* **52**:104–116.

41. **Lewis, G. K., and J. W. Goodman.** 1978. Purification of functional determinant-specific idiotype-bearing murine T cells. *J. Exp. Med.* **148**:915–924.

42. **Marrack, P., and J. Kappler.** 1986. The antigen-specific, MHC-restricted receptor on T cells. *Adv. Immunol.* **38**:1–30.

43. **McNamara, M., and H. Köhler.** 1984. Regulatory idiotypes. Induction of idiotype-recognizing helper T cells by free light and heavy chains. *J. Exp. Med.* **159**:623–628.

44. **Nisonoff, A., and E. Lamoyi.** 1981. Implications of the presence of an internal image of the antigen in anti-idiotypic antibodies: possible application to vaccine production. *Clin. Immunol. Immunopathol.* **21**:397–406.

45. **Novotny, J., M. Handschumacher, and E. Haber.** 1986. Location of antigenic epitopes on antibody molecules. *J. Mol. Biol.* **189**:715–721.

46. **Rajewsky, K., and T. Takemori.** 1983. Genetics, expression, and functions of idiotypes. *Annu. Rev. Immunol.* **1**:569–607.

47. **Rodkey, L. S.** 1980. Autoregulation of immune response via idiotype network interactions. *Microbiol. Rev.* **44**:631–659.

48. **Rudikoff, S.** 1983. Immunoglobulin structure-function correlates: antigen binding and idiotypes. *Contemp. Top. Mol. Immunol.* **9**:169–209.

49. **Sacki, Y., J. J. Chen, L. Shi, S. Raychaudhuri, and H. Köhler.** 1989. Characterization of "regulatory" idiotope-specific T cell clones to a monoclonal anti-idiotypic antibody mimicking a tumor-associated antigen (TAA). *J. Immunol.* **142**:1046–1052.

50. **Sacks, D. L., G. H. Kelsoe, and D. H. Sachs.** 1983. Induction of immune responses with anti-idiotypic antibodies: implications for the induction of protective immunity. *Springer Semin. Immunopathol.* **6**:79–97.

51. **Sakato, N., M. Semma, H. N. Eisen, and T. Azuma.** 1982. A small hypervariable segment in the variable domain of an immunoglobulin light chain stimulates formation of anti-idiotypic suppressor cells. *Proc. Natl. Acad. Sci. USA* **79:**5396–5400.

52. **Sim, G.-K., I. A. MacVail, and A. A. Augustin.** 1986. T helper cell receptors: idiotypes and repertoire. *Immunol. Rev.* **90:**49–72.

53. **Singhai, R., and J. G. Levy.** 1987. Isolation of a T cell clone that reacts with both antigen and anti-idiotype: evidence for anti-idiotype as internal image for antigen at the T cell level. *Proc. Natl. Acad. Sci. USA* **84:**3836–3840.

54. **Stein, K. E., and T. Soderstrom.** 1984. Neonatal administration of idiotype or anti-idiotype primes for protection against E. coli K13 infection in mice. *J. Exp. Med.* **160:**1001–1011.

55. **Strayer, D. S., W. M. F. Lee, D. Rowley, and H. Köhler.** 1975. Anti-receptor antibody. II. Induction of long-term unresponsiveness in neonatal mice. *J. Immunol.* **114:**728–733.

56. **Suciu-Foca, N., C. Rohowsky-Kochan, E. Reed, R. Haars, V. Bonagura, D. W. King, and K. Reemstma.** 1985. Idiotypic network regulations of immune responses to HLA. *Fed. Proc.* **44:**2483–2487.

57. **Taniguchi, M., I. Takei, T. Sumida, M. Kanno, M. Tagawa, and T. Ito.** 1984. Suppressor T-cell hybridoma with a receptor recognizing KLH-specific suppressor factor, p. 435–447. *In* M. I. Greene and A. Nisonoff (ed.), *The Biology of Idiotypes.* Plenum Publishing Corp., New York.

58. **Tasiaux, N., R. Leeuwenkroon, C. Bruyns, and J. Urbain.** 1978. Possible occurrence and meaning of lymphocytes bearing auto-anti-idiotypic receptors during the immune response. *Eur. J. Immunol.* **8:**464–468.

59. **Trenkner, E., and R. Riblet.** 1975. Induction of anti-phosphorylcholine antibody formation by antiidiotypic antibodies *J. Exp. Med.* **142:**1121–1132.

60. **Urbain, J., C. Collignon, J. D. Franssen, B. Mariame, O. Leo, G. Urbain-Vansanten, P. Vandewalle, M. Wikler, and C Wuilmart.** 1979. Idiotypic network and self-recognition in the immune system. *Ann. Inst. Pasteur Immunol. Sect. C* **130:**281–288.

61. **Urbain, J., M. Wikler, J. D. Franssen, and C. Collignon.** 1977. Idiotypic regulation of the immune system by the induction of antibodies against anti-idiotypic antibodies. *Proc. Natl. Acad. Sci. USA* **74:**5126–5130.

62. **Waters, S. J., and C. Bona.** 1988. Characterization of a T-cell clone recognizing idiotypes as tumor-associated antigens. *Cell. Immunol.* **111:**87–93.

63. **Weinberger, J. F., R. N. Germain, S.-T. Ju, M. I. Greene, B. Benacerraf, and M. E. Dorf.** 1979. Hapten-specific T cell responses to 4-hydroxy-3-nitrophenyl acetyl. II. Demonstration of idiotypic determinants on suppressor T cells. *J. Exp. Med.* **150:**761–776.

64. **Weiss, S., and B. Bogen.** 1989. B lymphoma cells process and present their endogenous immunoglobulin to major histocompatibility complex-restricted T cells. *Proc. Natl. Acad. Sci. USA* **86:**282–286.

65. **Wikler, M., C. Demeur, G Dewasme, and J. Urbain.** 1980. Immunoregulatory role of maternal idiotopes. Ontogeny of immune networks. *J. Exp. Med.* **152:**1024–1035.

66. **Wilson, D. B., and D. L. Bellgrau.** 1982. Speculations on the nature of allospecific T-cell receptors and mechanism of tolerance to self-MHC gene products. *Behring Inst. Mitt.* **70:**210–212.

67. **Wright, A., J. E. Lee, M. P. Link, S. D. Smith, W. Carroll, R. Levy, C. Clayberger, and A. M. Kerensky.** 1989. Cytotoxic T lymphocytes specific for self tumor immunoglobulin express T cell receptor γ chain. *J. Exp. Med.* **169:**1557–1565.

Chapter 3

Viral Infections

Glen N. Gaulton and David B. Weiner

This chapter focuses on the origin, maturation, and experimental manipulation of the idiotypic network in immune responses to viruses. The theoretical considerations of the idiotypic network have been examined by Weido et al. and Cerny and Hiernaux (this volume). Our presentation is limited to a distinction between anti-idiotopes (Ab_2) that recognize public idiotopes, private idiotopes, or internal-image idiotopes. Cross-reactive, public, or recurrent idiotopes are those common to multiple immunoglobulins within an individual and among most individuals of a species. These are more prevalent in antibodies that share specificity, but this is not absolute. In some cases interspecies public idiotopes may be found. This probably represents the use of a particular V_H or V_L gene family. In keeping with previous nomenclature, these are termed IdX (38, 54). Private idiotopes, termed IdI, are those created by unique combinatorial association among variable (V) regions or sequence variation within a V region. These are largely unique to individual antibody responses within individuals. The distinction of internal-image anti-idiotopes (IdB) is empirical. Internal-image anti-idiotopes bear an idiotope that is complementary to the paratope of Ab_1. Thus, internal images show no species restriction. These different forms of anti-idiotopes will be operationally defined in the following sections. We will use the terms idiotope when referring to monoclonal antibodies or when epitope specificity has been established and idiotype when referring to polyclonal responses or when epitope specificity is ambiguous.

Our initial discussion will center on the nature of the idiotypic repertoire in primary and secondary antivirus antibody responses. These obser-

Glen N. Gaulton • Department of Pathology and Laboratory Medicine, University of Pennsylvania School of Medicine, Philadelphia, Pennsylvania 19104.   **David B. Weiner** • Department of Pathology and Laboratory Medicine and Department of Medicine, University of Pennsylvania School of Medicine, Philadelphia, Pennsylvania 19104.

vations are fairly limited, and so we will restrict our treatment to observations made in murine and human systems following exposure to either influenza virus or hepatitis B virus. We next present the rather extensive evidence on manipulation of both B- and T-cell immunity to virus by using anti-idiotypic reagents. The success of these manipulations clearly depends on the extent of idiotypic dominance or cross-reactivity of antivirus responses. This discussion will focus on the hepatitis virus, mammalian reovirus, and Sendai virus systems as representative examples of vaccine development strategies. We also include a brief discussion of the current data and potential applications of these reagents as acquired immunodeficiency syndrome (AIDS) virus vaccines. Internal-image anti-idiotypic antibodies have also been widely used as tools to probe the interaction of viruses and cellular targets, particularly at the level of cellular attachment sites. Applications of these techniques to the dissection of the pathogenicity of reovirus and Semliki Forest virus will also be explored. Finally, we will conclude with a discussion of the relationship of idiotypic networks, virus mimicry, and internal-image autoantibody production in disease.

PARTICIPATION OF THE IDIOTYPIC NETWORK IN IMMUNE RESPONSES TO VIRUS

Evidence for Restricted Idiotypes in Antivirus Responses

The prediction by Jerne (54) that the immune system is self-regulated by a network of idiotypic–anti-idiotypic interactions was quickly followed by the first experimental demonstration of auto-anti-idiotypic antibodies against the T15 idiotope in BALB/c mice hyperimmunized with *Streptococcus pneumoniae* (68). Kluskens and Kohler (68) documented both the presence and specificity of anti-T15 idiotope by using T15-coated sheep erythrocytes. This observation was succeeded by numerous reports of auto-anti-idiotypes produced in the normal course of immune responses. These include such diverse antigens as tetanus toxoid, dextran, levan, sheep erythrocytes, major histocompatibility complex (MHC) antigens, numerous haptens, and antibody molecules themselves (reviewed in reference 119).

The first direct demonstration of an endogenous anti-idiotypic component in an antivirus immune response was not made until 1985, when Troisi and Hollinger (120) were able to detect anti-anti-hepatitis B surface antigens (HBsAg) in the sera of both acute and chronic hepatitis B-infected patients by radioimmunoassay with anti-immunoglobulin M (IgM)-coated beads and radiolabeled anti-HBsAg. The participation of the idiotypic network in the regulation of antivirus responses was, however, reliably predicted far in advance of this report. The first clues to this interaction were uncovered by Liu et al. (77) following the observation of shared idiotopes in antibodies reactive with different determinants of the influenza virus hemagglutinin (HA). Monoclonal antibodies were prepared to influenza viruses A/PR/8/34 (PR8) and B/Lee/40 (B/Lee), and unique reactivity patterns were defined by using virus variants. Cross-reactive idiotypes (IdX) were detected by binding of an anti-idiotypic antiserum to three different anti-PR8 monoclonal antibodies that displayed different epitope specificity. Binding inhibition indicated that these idiotopes were closely associated with the anti-PR8 paratope.

A more refined analysis of the IdI and IdX components of this response was performed by using anti-B/Lee monoclonal antibodies. IdI components were studied by using syngeneic antiserum to individual monoclones. One IdI, defined by anti-B147, was unique (one of seven monoclones positive). The others tested (e.g., anti-B118) had a limited distribution (two of seven monoclones positive). IdX were detected in all seven monoclones by using an HA inhibition assay; however, IdX for B/Lee were not reactive with PR8. The distribution of IdX in primary and secondary virus humoral immune responses to B/Lee was also examined. Two IdX were present during peak primary and secondary responses at equivalent levels. One IdX, however, was observed only during the primary response, whereas another was dominant only in the secondary response. Interestingly, in the latter instance the majority of IdX (89%) was resident in antibody that did not bind B/Lee virus. Auto-anti-IdX was not detected in these animals. In a subsequent report, Moran et al. (83) observed a sharing of IdX by monoclonal antibodies reactive with both PR8 and the unrelated X-31 influenza virus. In one unique case an X-31-reactive antibody that shared PR8 IdX was derived from a mouse that had been immunized with only PR8.

These observations document the presence of multiple families of idiotypic determinants in the immune response to influenza virus. Some idiotopes are unique, others display a clonal bias, and many are broadly cross-reactive. Differential expression of these determinants may be the consequence of the preferential expression of V_H, diversity (D), joining (J), or V_L genes, the expansion of selected clones through somatic mutation and antigen selection, and/or clonal amplification through the actions of idiotype-specific regulatory cells.

The following discussion should help to resolve some of these issues. This presentation will be largely restricted to immunoglobulin idiotypes. Observation of T-cell receptor idiotypes in antiviral responses is limited to reports of restricted heterogeneity in the idiotypy of the cytolytic response to Newcastle disease virus (56). Regulatory aspects of T-cell receptor idiotypes will be presented in Experimental Manipulation of the Idiotypic Network in the Regulation of Immune Responses to Virus (see below).

Molecular Events That Regulate Idiotope Expression

The first attempt at a molecular description of the events that control restricted or biased idiotope expression in the antivirus response was begun by Staudt and Gerhard (110), working with the anti-influenza (PR8) response. These investigators observed a biased light-chain usage in the anti-PR8 response of the total murine population and restricted (but not identical) idiotypic patterns by individual members within the population. A panel of 125 anti-HA hybridomas were isolated from the secondary response to PR8 of 14 BALB/c mice. Antibodies displayed 104 distinct reactivity patterns (clonotypes) when tested on 39 mutant PR8 viruses and 12 epidemic strains. This response defined four specificity groups to four antigenic sites on the HA, designated Sa, Sb, Ca, and Cb (44). The predicted size of the anti-HA repertoire was 1,500 clonotypes.

Idiotypic analysis of hybridomas isolated from individual mice indicated a clear idiotypic bias within each mouse. For example, in mouse 37, 27 of 54 hybridomas (50%) had a common idiotype, yet this idiotype was present in only 1.4% of hybridomas in the total population. The responses of individual mice also yielded paratypically related antibodies; 86% of Id$^+$ antibodies from mouse 37 failed to bind a panel of Sb mutants. Despite these similarities, fine specificity analysis showed 18 distinct reactivity patterns and idiotypic microheterogeneity (IdI) after absorption to remove anti-constant-region specificities.

An analysis of V_L use within these antibodies showed that 22% of the anti-HA panel used members of the V_K21 group, compared with only 8.6% use in normal BALB/c serum immunoglobulin. Interestingly, in mouse 36 all members of the H36-5Id family used V_K21C (IdX). The most probable explanation for these results is that the idiotypic families detected descended from common precursor cells by somatic mutation.

This hypothesis was confirmed and extended by McKean et al. (82). Sequence analysis was performed on the amino-terminal residues of seven antibodies isolated from mouse 36 (described above). All seven were members of the V_K21C family and bound to the same antigenic region, Sb, on the HA molecule. Members could, however, be distinguished by IdI analysis and/or reactivity patterns within Sb. Sequence analysis confirmed that all of the antibodies belong to the V_K21C subgroup and that all were rearranged to the J_K2 gene segment. Amino acid replacements among H36 members were remarkable for both their magnitude and location. The majority of antibodies contained more than replacements, most of which were parallel clusters in the CDR1 region. Sequence analysis of mRNA from these hybridomas supported these results and allowed assessment of the relative contribution of germ line diversity to the anti-HA response (13). Although combinatorial joining and association contribute to diversity, the fine specificity of antibodies in these mice appears to be linked primarily to sequential somatic mutation. Analysis of several other independently derived and selected anti-HA antibodies has pointed out that restricted V_H gene usage (8) and combinatorial joining and association (94) can also play important roles in regulating idiotypic dominance and paratope specificity.

Demonstration of Dominant Idiotopes in Human Antiviral Responses

The phenomenon of idiotypic dominance has also been observed in human antiviral responses. Kennedy and Dreesman (60) detected a common human anti-idiotype in purified anti-HBsAg isolated from nine individuals, but not in normal immunoglobulin, by using rabbit anti-anti-HBsAg sera. Anti-idiotype–idiotype binding was inhibited by both HBsAg and a virus-derived HBsAg polypeptide. Thus, the idiotype was judged to be associated with the antibody-combining site. Further characterization of this common anti-idiotype revealed that it was induced by the group *a* determinant. Three different HBsAg preparations purified from three pools of human plasma positive for HBsAg *adw, ayw,* or *adr* subtypes inhibited the anti-idiotype–idiotype reaction equally well (65). Anti-idiotype binding was also partially inhibited by a synthetic peptide corresponding to the major polypeptide (P25) of HBsAg (52), suggesting that this sequence is related to the epitope recognized by idiotype-positive anti-HBsAg.

An interspecies idiotype cross-reaction was detected on the anti-HBsAg produced in rabbits, guinea pigs, swine, goats, chimpanzees, and mice, but not chickens (63). These antibodies partially inhibited the binding of human anti-HBsAg to a rabbit anti-idiotype serum. Interestingly, there was no obvious correlation between total anti-HBsAg levels and inhibitory capacity. These observations suggest that the ability to respond to a particular HBsAg epitope is highly conserved in mammalian but not avian species. This may result from a dominant V_H or V_L gene use, as observed in the anti-PR8 response. Similar observations have recently been made by Sigal et al. in a comparison of the human and murine anti-influenza responses (106). These workers demonstrated that there was a striking predominance of lambda light chains in the idiotype-positive human response to influenza virus.

Linkage of idiotypic dominance to epitope recognition is further supported by the demonstration that only monoclonal anti-HBsAg that recognize the epitope defined by the cyclic synthetic peptide described above bear this interspecies idiotype (57). Despite this, partial inhibition by anti-HBsAg reflects the presence of a complex family of combining site-related idiotopes in mammals. A more refined dissection of this idiotypic pattern, conducted by Thanavala et al., verified the presence of a major IdX in the murine response to HBsAg (118).

Detection of Auto-Anti-Idiotypes following Virus Infection

As mentioned above, the production of virus-specific auto-anti-idio-types was first found in the sera of 67 of 140 (47.9%) hepatitis B virus-infected patients (120). A more pronounced effect was seen when sera were subdivided into HBsAg- or anti-HBs-positive groups. Approximately 93% of anti-idiotype-positive sera contained HBsAg and, correspondingly, 70% of HBsAg-positive sera contained anti-idiotype. These sera were negative for the presence of rheumatoid factor. Only 9.1% of anti-HBs-positive sera and 14.3% of anti-HBs/HBsAg-negative sera contained anti-idiotype. When these sera were analyzed for interspecies reactivity by using the reagents described above (52), inhibition was seen only with the mammalian anti-HBsAg and not the chicken anti-HBsAg. Anti-idiotype levels during acute and chronic hepatitis infection were investigated by Rath et al. (91). In both instances the anti-idiotype responses correlated with poor humoral and cellular responses to HBsAg.

A potential regulatory role of these anti-idiotypes was supported by the observation that anti-idiotype titers dropped in patients who lost hepatitis virus DNA or antigen in the serum (50). The most likely mechanistic explanations of these effects are that anti-idiotypic antibodies can suppress the synthesis of anti-HBs antibodies by their action on clones that bear the idiotype and/or that the presence of the anti-idiotype permits more active replication of virus by removing anti-HBs from the serum.

More detailed information on the regulatory roles of antivirus anti-idiotypes has been obtained by manipulation of immunity through the in vivo administration of idiotype or anti-idiotype. For example, injection with anti-idiotype in the HBsAg system has been shown to increase both the number of anti-HBsAg-secreting B cells (59) and the anti-HBsAg humoral titer (61, 64). As expected, the newly formed anti-HBsAg uniformly express the interspecies idiotype. These and other related observations have served as the impetus for the use of anti-idiotypes as vaccines and receptor probes. These topics provide the focus of the rest of the chapter.

EXPERIMENTAL MANIPULATION OF THE IDIOTYPIC NETWORK IN THE REGULATION OF IMMUNE RESPONSES TO VIRUS

Regulation of Idiotypic and Anti-Idiotypic Repertoires—the Principle of Idiotypic Vaccines

The preceding discussion emphasized the idiotypic bias of antivirus responses and provided some examples of the self-regulatory capacity of the idiotypic network. These two facts lie at the heart of our presentation of anti-idiotype-based vaccines. The inherent regulatory pathways of the idiotypic network have been most fully explored in several hapten-based model systems (reviewed by Cerny and Hiernaux [this volume]). There are numerous reports of the effects of anti-idiotype treatment on immune responses. These include demonstrations of both suppression and enhancement of idiotype expression (9, 18) and the induction of idiotype-specific suppressor (5) and helper (133) T-cell subsets.

The presence of both IdX and IdI elements in the auto-anti-idiotype response to virus predicts that administration of anti-idiotype (Ab_2) can alter immune responses in several ways. Ab_2 injection will induce several classes of anti-anti-idiotopes (Ab_3) that may either share (Ab_1') or lack antigen-binding capacity. Most observations indicate that Ab_1' constitute a minority of the Ab_3 response to a polyclonal Ab_2 immunogen (122). The induction of Ab_3 may arise from one of two pathways: Ab_2 may activate silent or Ab_1-related clones by virtue of shared or cross-reactive idiotopes, or it may function as internal-image Idβ. Definition of the characteristics of Idβ is based on the prediction that Idβ bear a region that mimics the three-dimensional structure of the epitope recognized by Ab_1. These characteristics are as follows: (i) Idβ should display the same reactivity pattern and affinity as antigen to Ab_1; (ii) Idβ should induce the same immune response as antigen, including selected clones, T-cell dependence, and species specificity; and (iii) Idβ should mimic the physiological functions of the epitope, for example, hormone signaling, that are unrelated to immunologic functions.

It is important to remember that Ab_2 may also have direct regulatory effects on Ab_1 clones irrespective of Ab_3 production. Alterations in the immune response following injection of these Ab_2 probably occur through direct interaction, or perturbation, of Ab_2 with idiotypes on surface immunoglobulin or T-cell receptors (6). As discussed above, these interactions may either enhance or suppress immunity. The parameters that govern suppression versus enhancement of immune responses to these Ab_2 have not been carefully defined.

These observations have led investigators to the conclusion that Idβ are the most reliable and reproducible reagents for the induction of immunity. When possible, these have indeed been the reagents of experimental choice, although there is a substantial literature on the effects on IdX and IdI anti-idiotypes as vaccines.

Advantages and Disadvantages of Anti-Idiotypic Vaccines

There are several attractive advantages of anti-idiotype-based vaccines. These relate to the availability and infectious nature of conventional antigen, genetic and ontogenic restrictions of hosts, and repertoire selection.

Potential difficulties include anaphylactic reactions, a more restricted epitope response repertoire, and the management of long-lasting, high-titer responses.

Conventional vaccines are not suitable in many instances owing to the infectious nature of the organism or toxic side effects. Examples include the highly infectious agents of leprosy and protozoal diseases. Difficulties also arise in the use of attenuated viral vaccines that contain large concentrations of genomic material with the potential for reversion to a virulent form or transformation of host cells. This is particularly true for members of the herpesvirus and human retrovirus families. The use of purified, synthetic, or recombinant proteins as immunogens seems to be a good alternative; however, these methods are currently plagued by high cost, side effects of expression vectors, and poor immunogenicity (33). Anti-idiotype vaccines are noninfectious, free of toxoids, and easy to purify and administer.

Anti-idiotype vaccines, by their nature, stimulate a more restricted immune response than do either whole organisms or purified peptides. One advantage of this is that potentially damaging autoimmune reactions of a complex immunogen may be avoided. This restriction may, however, have severe disadvantages. Responses directed to one or a few epitopes of an organism are more prone to antigenic drift or genetically linked nonresponsiveness. As an example of the former, one challenge of developing an effective vaccine to human immunodeficiency virus (HIV) is that of defining antigenic regions of the envelope that do not exhibit high mutation rates. A clear example of the genetic restriction of anti-idiotype immunity is provided by the strain dependence of resistance to *Trypanosoma rhodesiense* (98). Failure to respond to anti-idiotype immunization in these systems probably results from the lack of IdX or IdI in the genome of selected strains. Problems with genetic restriction are largely overcome with internal-image-based vaccines. IdB responses display the same restriction pattern as antigen (epitope). Indeed, one of the distinguishing characteristics of Idβ is the ability to induce immunity across species barriers.

The nature of the immune response induced by anti-idiotype may differ significantly from that of antigen. Presumably, this relates to the different way that the immune system sees antigen or Idβ compared with IdX or IdI. For example, whereas Idβ may be processed and presented in an antigenlike MHC-restricted fashion, IdI and most certainly the majority of IdX can directly interact with lymphocyte clones, thereby altering immune responses through network perturbation. One manifestation of this inherent difference is the induction of silent antigen-binding clones by anti-idiotype (4, 36). This might prove useful for the induction of immunity in certain situations of immune incompetence. As an example, Stein and Soderstrom (111) and Gaulton et al. (42) have demonstrated that anti-idiotypic immunization can provide an effective barrier to potentially lethal neonatal infections.

A sequela of these observations is that the regulation of immune-response strength and duration is more variable with anti-idiotype vaccines than antigen. Immunization with live virus typically generates prolonged humoral and cellular responses. Immune responses to attenuated viruses and peptides often decline rapidly after antigen clearance (33). Both high-titer and prolonged responses have been induced by using anti-idiotypes (42, 62); however, this usually requires multiple injections in adjuvant and coupling to a carrier molecule. As mentioned above, the injection of Ab_2 does not always lead to increased immunity. Suppression of immune responses can be a frequent result. The physical state of Ab_2 and route of injection can have important effects on the outcome (48, 58, 74).

The most serious problem with anti-idiotype vaccines is the repeated administration of heterologous antibody. Even the development of human hybridomas for use in humans will not escape potentially serious allogeneic, allergic, and immune-complex reactions when multiple injections are used. The most promising avenue is that molecular dissection of internal-image structure on Ab_2 may allow the development of synthetic peptides that mimic this complementarity. This approach has been exploited by Greene and colleagues, using the reovirus system (7, 130).

These observations illustrate both the benefits and pitfalls of anti-idiotype vaccines. Specific examples of how these properties have been manipulated to confer immunity and protection to virus infection are presented below.

Examples of Anti-Idiotypic Vaccine Deployment

The first studies that used anti-idiotypic antibodies in an experimental setting for the induction of antivirus immune responses were those of Urbain et al. (121, 122) with tobacco mosaic virus (TMV). Rabbit anti-idiotype prepared to anti-TMV detected a shared idiotype on anti-TMV of rabbits, mice, horses, goats, and chickens. This anti-idiotype induced the formation of anti-TMV when injected into mice, without the need for a subsequent antigen boost. Thus, from a strictly immunologic perspective, this has all appearances of an internal-image anti-idiotype. In an interesting follow-up to this work, Francotte and Urbain (36) demonstrated a similar induction of anti-TMV by using a rabbit anti-idiotype that recognized an idiotope that was presumably nonrecurrent across species. This induction of silent clones most probably reflects a regulatory idiotypic control. These studies formed the basis for a multitude of related investigations using anti-idiotypic tools. The three most prominent systems are (i) clinical application of antivirus vaccines (hepatitis B virus), (ii) examination of the structural basis of internal image (reovirus), and (iii) effectiveness of T-cell receptor anti-idiotypes (Sendai virus).

Development of an anti-hepatitis B virus anti-idiotype vaccine

The human immune response to hepatitis B virus contains a common idiotype with a broad species distribution. Evidence for this was presented in Participation of the Idiotypic Network in Immune Responses to Virus (see above). This idiotype was detected in 30 of 32 individuals who were either naturally infected or immunized with HBsAg (57). The idiotype is present on antibodies to a conformationally dependent group *a* determinant and is combining-site related (60, 65). Anti-idiotypes to this common anti-HBs idiotype have been extensively used to manipulate immune responses to the hepatitis B virus. Anti-idiotypes administered in combination with HBsAg or alone were shown to augment the number of anti-HBs-secreting spleen cells and the level of IgM in mouse serum (59, 61). Similar results were obtained when anti-idiotype was used in combination with synthetic cyclic HBs peptides (66). Anti-HBs levels in these mice were considerably lower than in mice that received anti-idiotype and intact HBsAg; however, they were equivalent to the response to a single injection of HBsAg. Interestingly, mice that received two injections of anti-idiotype alone or in combination with peptide were found to express the common idiotype on anti-HBs, and this response was selective for the group-specific *a* determinant (59, 64).

The induction of a group-specific *a* response suggests that the anti-idiotype is acting as a serological mimic of the HBsAg, i.e., an internal image. To solidify this hypothesis, rabbits used to prepare anti-idiotype were rested for 14 months and then reimmunized with autologous anti-idiotype. Prior to immunization, no anti-HBs were detected, although there was still a weak anti-human immunoglobulin response. After four 500-μg injections of anti-idiotype in alum with adjuvant, strong anti-HBs responses were seen (57).

The more important observation from the standpoint of vaccine development is that the group-specific *a* response was previously shown to be protective for infection with hepatitis B virus (113). Toward this end, techniques were optimized for the use of anti-idiotypes in primates. Increased antibody titers and IgG isotypes were induced when anti-idiotype was precipitated in alum (61). Two chimpanzees were immunized with four 1-mg doses of anti-idiotype in alum over 11 weeks (62). By week 12, anti-HBs responses were detected above control chimpanzee sera, and by week 20, the responses were >1,000-fold above background levels. The anti-HBs response was verified to be Ab_3 by selective binding of the rabbit anti-idiotype (Ab_2) after absorption of xenogeneic reactivity.

To demonstrate a protective effect, we challenged chimpanzees at week

24 with 3,000 infectious doses of hepatitis B virus. The two chimpanzees that received anti-idiotype were negative for all markers of hepatitis B virus infection out to week 70. These analyses were performed with serologic, histologic, and molecular probes for virus. By all criteria, these animals were protected from hepatitis infection. In contrast, two chimpanzees that received either control antibody or sham injection developed hepatitis B infection within 4 weeks (week 28) after challenge, with infectious particles detected after 7 weeks (week 31).

These studies answer several questions regarding the effectiveness of anti-idiotype vaccines. The first is the longevity of immune responses. In the chimpanzee studies, anti-HBs levels plateaued at approximately 13 weeks after the last anti-idiotype injection. These levels were maintained out to week 70. Thus, the anti-HBs response is not transient. The titer of responses in chimpanzees can be compared only with those in one animal that was injected with a single dose of hepatitis virus. Anti-HBs titers were 1,000-fold and 200-fold greater than the anti-hepatitis B virus response in the two animals injected with anti-idiotype. In previous studies, titers induced by anti-idiotype alone were six- to ninefold lower than those induced by a combination of anti-idiotype and HBsAg (61).

Concerns about the consequences of multiple injections of heterologous antibody were also addressed by Kennedy et al. (62). All of the animals used for these studies were healthy, despite receiving a total of 4 mg of rabbit IgG over an 11-week interval. All animals developed a serum anti-rabbit IgG response that was IgG in nature. Skin testing with alum-precipitated rabbit anti-idiotype failed to demonstrate any allergic reaction.

The development and use of monoclonal anti-idiotopes to anti-HBs have been examined by Thanavala et al. and Colucci et al. Thanavala et al. (116) used a classical approach to the development of anti-idiotopes. Anti-idiotopes were obtained by immunization of mice with a monoclonal anti-HBsAg that recognizes a conserved epitope of HBsAg. Two monoclonal anti-idiotypic antibodies reacted with anti-HBs from four different species and, thus, showed properties of internal image. An analysis of the affinity and extent of reactivity of anti-idiotopes and HBsAg with several preparations of anti-HBs allowed a prediction of the diversity and efficacy of anti-idiotopes as vaccines (117). Anti-idiotypic antibodies uniformly displayed higher affinity for anti-HBs immunoglobulin than did either synthetic HBs peptide or a gp30p25 polypeptide complex that exhibits all the antigenic repertoire of native HBsAg. However, peptides bound 10 to 20 times more anti-HBs on a molar basis than did anti-idiotope. These results indicate that immunization with anti-idiotope will elicit a higher-affinity, although more restricted, anti-HBs response than obtained with peptide or HBsAg immunization. These workers concluded, in agreement with Kennedy et al. (62,

64, 66), that the most efficacious vaccine would comprise a combination of peptide and anti-idiotope, or a cluster of anti-idiotopes.

The approach taken by Colucci et al. (16) for the development of monoclonal anti-idiotopes utilized the properties of molecular mimicry of internal-image antibodies (see Idiotypic Reagents as Tools for the Dissection of Virus Biology). Anti-idiotypes were prepared by immunization of mice with human antibodies to polymeric human serum albumin (poly-HSA). PolyHSA binds to both HBsAg and hepatocytes. Anti-polyHSA (Ab_1) were used to prepare monoclonal anti-idiotopes (Ab_2) in mice. These antibodies mimicked the binding of polyHSA to HBsAg (16). Syngeneic anti-anti-idiotopes (Ab_3) were then prepared and tested for immunologic mimicry of HBsAg. Anti-anti-idiotopes induced a potent anti-HBs response in rabbits that was equivalent to injection with HBsAg. Monclonality and epitope specificity have not yet been demonstrated in this system.

Examination of the structural basis of internal image

The structural basis of internal image has been most thoroughly examined for anti-idiotopes to reovirus type 3. The production, purification, and screening of polyclonal reovirus type 3-specific anti-idiotypic antibodies have been described by Nepom et al. (85). Briefly, immunoglobulins from mice immunized with type 3 virus were first absorbed with type 1 virus and the reassortant virus 3HA1 (a type 3 virus with a type 1 HA). This removed all antibodies not reactive with the type 3 HA. Immunoglobulins were then absorbed with preimmune mouse immunoglobulin to remove xenogeneic reactive antibody. Rabbits were immunized with these polyclonal anti-HA3-specific antibodies, and the presence of anti-HA3 anti-idiotypes in the resultant sera was verified by radioimmunoassay. A panel of monoclonal anti-HA3-specific antibodies was then screened for the ability to inhibit the binding of type 3 anti-idiotype to rabbit anti-HA3. One neutralizing monoclonal antibody, termed 9BG5, inhibited 83% of anti-idiotype binding, indicating that this epitope is immunodominant in the type 3 anti-idiotype response.

On the basis of this finding, 9BG5 immunoglobulin was used as an immunoabsorbent to affinity purify HA3-specific anti-idiotypes from crude immunoglobulin fractions. Indirect-immunofluorescence staining with purified anti-idiotype indicated that the pattern of anti-idiotype binding mimicked the tissue tropism of reovirus type 3 but not type 1 (84, 85). A monoclonal equivalent of this internal-image behavior was isolated by Greene and colleagues (55, 86). BALB/c mice were injected with the syngeneic antibody 9BG5, and anti-HA3 specific anti-idiotopes were identified by the ability to inhibit the binding of 9BG5 to HA3 and by inhibition of polyclonal anti-idiotype binding to rabbit anti-HA3 (85, 86). Characterization of HA3 mimetic binding properties of the monoclonal anti-idiotope 87.92.6 antibody was initially described by Kauffman et al. (55). By all physical and functional criteria tested to date, the 87.92.6 monoclonal antibody acts as an internal image of the HA3 molecule. These include identical cell-binding patterns, inhibition of virus binding, inhibition of virus infectivity, and mimicry of virus receptor-dependent changes in cellular physiology (39). A more detailed treatment of the biochemical aspects of virus receptor studies conducted with these antibodies is presented below (Experimental Applications of Anti-Idiotypic Probes).

An important correlate to the demonstration of biochemical mimicry is that of immunologic mimicry as described above. The naïve prediction is that internal image should stimulate both T- and B-cell immunity in a manner analogous to that of epitope. The network theory predicts that immunization with internal-image anti-idiotypic antibody will elicit an Ab_3 response that will contain a subset of Ab_1-like (Ab_1') antibodies. These Ab_1' antibodies display the same epitope specificity as Ab_1. Thus, stimulation of B cells by internal-image Ab_2 is expected, and routinely observed. This could occur through recognition of monovalent internal-image epitopes on Ab_2 (à la antigen), or by polyvalent clustering of surface immunoglobulin via idiotope recognition (à la regulatory idiotopes). There is also evidence that T cells can be stimulated directly by cross-linking of regulatory idiotopes shared with immunoglobulin (46, 51). However, stimulation of T-cell function through MHC-restricted presentation of the internal image is more complicated. Because presentation of antigen relies on the generation of small antigenic fragments that may contain limited secondary structure, it seems likely that internal images that reflect the primary structure of related epitopes, in contrast to those that represent conformational images, are more likely to mimic MHC-restricted T-cell function.

Reovirus HA3 is a potent inducer of serotype-specific T- and B-cell immunity (32, 85). Immunization with 87.92.6 anti-idiotope showed identical MHC-restricted and serotype-specific HA3 immunity (40, 42, 104, 105). Immunization of syngeneic mice with purified anti-idiotope was shown to stimulate potent cytolytic, delayed-type hypersensitivity (DTH), and helper T-cell responses. Potent and transferrable DTH responses, detected by the footpad-swelling assay, were detected by 5 days after subcutaneous immunization with 100 µg of 87.92.6 in saline. Maximal swelling was seen at 28 to 32 h after challenge with virus, and footpads displayed extensive invasion of mononuclear cells on morphological analysis. The magnitude of this response was approximately 70% of that seen in animals immunized with type 3 virus.

The specificity of DTH responses to the HA3 protein is demonstrated in Table 1 (105). Significant swelling was detected only after challenge with

Table 1. Specificity of Syngeneic DTH Responses Elicited by
Reovirus Anti-Idiotope[a]

Immunization	Challenge	DTH response $(10^{-2}$ mm$)$[b]
Type 3	Type 3	37.8 ± 2.3[c]
Anti-idiotope	Type 3	25.5 ± 1.7[c]
Anti-idiotope	Type 1	8.5 ± 1.5
Anti-idiotope	1HA3	27.3 ± 2.0[c]
Anti-idiotope	3HA1	9.5 ± 1.6
Anti-idiotope	Variant K	11.8 ± 1.7
Saline	Type 3	8.5 ± 1.5

[a] Adapted from reference 105.
[b] DTH responses are expressed as the mean footpad swelling 24 h after challenge.
[c] $P < 0.005$ ($n = 6$) relative to saline control by two-tailed Student's t test.

type 3 virus (HA3) or the reassortant virus 1HA3 (type 3 HA on a type 1
background). Responses were not detected after challenge with either type 1
virus (HA1), the reassortant 3HA1, or the type 3 variant virus K (does not
bind Ab_1).

Attempts to induce cytolytic T-cell (CTL) responses by using a similar
protocol yielded interesting results. Injection of anti-idiotope in saline failed
to stimulate CTL activity; however, injection of 3×10^5 irradiated 87.92.6
hybridoma cells primed a CTL response of equal magnitude to that seen with
live-virus priming (104, 105). The serotype specificity and MHC restriction
of this response were not tested. However, a detailed survey of reovirus-specific CTL clones made in response to immunization with live virus indicated
that 25 to 40% of these clones recognized the 87.92.6 anti-idiotope (28, 104).
This determination was based on the ability of reovirus CTL lines prepared by
limiting dilution to lyse the 87.92.6 hybridoma in an MHC-restricted manner. A representative sample of this effect is presented in Table 2. Reovirus
CTL were isolated after mixed in vivo-in vitro stimulation, subjected to
limiting dilution, and assayed against the targets listed at an effector-to-target
cell ratio of 20:1. The results indicate that approximately 40% of the CTL
response to reovirus is directed to the idiotope defined by 87.92.6. When
effectors were generated from the MHC-incompatible strain B10.BR ($H-2^k$),
lysis of 87.92.6 ($H-2^d$) was not seen (28). Lysis was also inhibited by the
addition of antibodies to reovirus, H-2, LFA-1, or 87.92.6 immunoglobulin
(28, 104). Together, these results suggest, but do not prove, that anti-idiotope
recognition by T cells proceeds via MHC-restricted presentation or recognition, in contrast to direct T-cell receptor stimulation.

Our initial evidence for the induction of serotype-specific helper T cell

Table 2. Reovirus Type 3-Specific CTL Lyse Anti-Idiotope Hybridoma Cells[a]

Reovirus-specific CTL line	% Specific lysis of:		
	Unfected P815 cells	Type 3-infected P815 cells	87.92.6 hybridoma cells
1	6.4	40.0	21.0
2	8.3	43.0	29.9
3	8.9	38.0	31.7
4	4.6	29.8	22.7
5	12.2	39.0	24.4

[a] CTL assays were performed as described previously (104) by using a 5-h Cr release assay. (Adapted from reference 104.)

activity by anti-idiotope is shown in Table 3. BALB/c mice were injected on day 0 with either rabbit anti-idiotype (100 µg in complete Freund adjuvant), 2×10^7 type 3 reovirus particles, or saline. Thirty days later groups 3 and 4 were boosted with 1×10^5 type 3 particles, and all mice were bled on day 40. Neutralizing titer was determined by using a standard plaque assay method. Injection of anti-idiotype in adjuvant stimulated a weak response ($P < 0.01$; $n = 5$) compared with controls (group 1 versus group 5). A more pronounced effect was seen when anti-idiotype-primed mice were boosted with a subimmunogenic dose of reovirus (group 4). Under these conditions, antibody titers were equivalent ($P < 0.005$; $n = 5$) to those obtained with virus priming (group 2). As expected, the anti-idiotype response was serotype specific for HA3.

On the basis of these observations, a detailed study was conducted to evaluate the carrier, adjuvant, dosage, and kinetic requirements of immunization with monoclonal anti-idiotope (42). Preliminary analyses indicated

Table 3. Induction of Type 3-Specific Neutralizing Antibody with Anti-Idiotope[a]

Group	Immunization		Neutralization titer[b]	
	1	2	Type 3	Type 1
1	Anti-idiotype		1,500	<50
2	Reovirus type 3[c]		4,000	500
3		Reovirus type 3[d]	<50	<50
4	Anti-idiotype	Reovirus type 3[d]	3,500	<50
5			<50	<50

[a] From reference 40.
[b] Neutralization titer was determined by the viral plaque assay and represents the lowest dilution that yielded a 50% reduction in the plaque count.
[c] 2×10^7 type 3 particles.
[d] 1×10^5 type 3 particles.

that syngeneic anti-idiotope was not immunogeneic unless coupled to a carrier moiety. Multiple immunizations with up to 200 µg of anti-idiotope yielded neutralization titers only 5% of those obtained by immunization with virus or polyclonal anti-idiotype. In contrast, immunization with 25 µg of anti-idiotope coupled to keyhole limpet hemocyanin stimulated a strong anti-HA3-specific response. High-titer responses required multiple (at least five) immunizations. Dosage requirements were evaluated in the absence of adjuvant. Maximal responses were seen with >100 µg of anti-idiotope–keyhole limpet hemocyanin per injection. The concentration effect was remarkably similar to that seen for the induction of DTH responses (105). Immunization in the presence of either complete Freund adjuvant or alum adjuvant increased titers more than threefold. Under conditions that included multiple injections of keyhole limpet hemocyanin-coupled anti-idiotope (100 µg) in the presence of adjuvant, the anti-HA3 titers were approximately 75% of those observed after immunization with reovirus. Evidence for internal image was supported by the observation that injection of anti-idiotope stimulated anti-HA3 responses in four different mouse strains and across three species.

The more practical aspect of anti-idiotope vaccine development was evaluated by testing for prophylaxis. When injected into suckling mice, reovirus type 3 causes a fatal acute encephalitis within 8 to 10 days. Neonates are particularly susceptible, with a 50% lethal dose of 10^4 infectious viral particles administered on day 2 after birth. This provides a unique model for studying the prophylactic effects of passively transferred antivirus antibodies from mother to offspring and, more specifically, the use of maternal immunization with anti-idiotope as fetal prophylaxis. Females were first injected with anti-idiotope–keyhole limpet hemocyanin in alum or inactivated type 3 virus and then placed in mating cages. Animals were boosted biweekly until fertilization. To test for protection, weight-matched littermates were injected intramuscularly with reovirus type 3 or 1 on day 2 or 3 after birth and examined daily thereafter for neurologic abnormalities. The results of this study (Table 4) demonstrate that maternal immunization with anti-idiotope is effective in blocking all behavioral signs of reovirus infection and mortality. This effect was specific for the inhibition of type 3 virus infection. Because 2- to 3-day-old infants do not produce significant antibody responses, we assume that the protective effects are the results of passively transferred anti-HA3 (Ab_1') that bind to the neutralization domain of HA3 recognized by the 9BG5 (Ab_1) monoclonal antibody.

These results document the most extensive functional characterization of any anti-idiotope to date. The structural basis of this mimicry was addressed by direct sequence analysis and comparison of anti-idiotope 87.92.6 and HA3 (7). This revealed sequence similarity between amino acids 317 to

Table 4. Anti-Idiotope Immunization Confers Neonatal Protection[a]

Expt no.	Maternal treatment	Neonatal exposure to virus	No. of mice	% Abnormalities (day 10)	% Survival (day 15)
1	None	Type 3 (day 2)[b]	6	100	0
	Anti-idiotype	Type 3 (day 2)[b]	7	0	100
	Type 3 virus	Type 3 (day 2)[b]	6	0	100
	None	Type 1 (day 2)[c]	6	100	83
	Anti-idiotype	Type 1 (day 2)[c]	5	100	80
	Type 1 virus	Type 1 (day 2)[c]	5	0	100
2	None	Type 3 (day 3)[d]	6	100	33
	Anti-idiotype	Type 3 (day 3)[d]	6	0	100

[a] From reference 42.
[b] Challenged with 1×10^4 PFU of reovirus type 3 (10 50% lethal doses) intramuscularly on day 2.
[c] Challenged with 1×10^4 PFU of reovirus type 1 intramuscularly on day 2.
[d] Challenged with 5×10^4 PFU of reovirus type 3 (50 50% lethal doses) intramuscularly on day 3.

332 of the HA3 and a combined determinant composed of the second complementarity-determining regions (CDRII) of both the V_H (amino acid 43 to 56) and V_L (amino acids 39 to 55) residues of 87.92.6. These sequences are shown in Fig. 1. The similarity between residues 323 to 332 of HA3 and 46 to 55 of V_L is noticeable, with five identical and three conserved amino acids. This is particularly interesting, as the CDRs of immunoglobulins have been identified as antigen contact residues (49, 107). The Chou and Fasman algorithm was used to predict the secondary structure of this site. Both peptides were predicted to fold into reverse turns involving the sequence Tyr-Ser-Gly-Ser. These observations prompted Williams et al. to propose that the internal image activity of 87.92.6 was a consequence of the molecular structure of this region (130, 131). This hypothesis was examined by the construction and functional analysis of synthetic peptides derived from this region.

Four synthetic peptides were prepared for the evaluation of molecular mimicry. These were termed V_L peptide (containing the V_L CDRII), V_H

```
V_H    43  Gln Gly Leu Glu Trp Ilu Gly Arg Ilu Asp Pro Ala Asn Gly
            •   o   o       •   •   •   o
Reo   317  Gln Ser Met --- Trp Ilu Gly Ilu Val Ser Tyr Ser Gly Ser Gly Leu Asn
                               o   o           •   •   •   •       •   o
V_L    39  Lys Pro Gly Lys Thr Asn Lys Leu Leu Ilu Tyr Ser Gly Ser Thr Leu Gln
```

Figure 1. Amino acid sequence similarity between 87.92.6 and reovirus HA3. (Adapted from reference 7.)

Table 5. Presence of Neutralizing Antibody in Antipeptide Sera[a]

Immunization	50% Neutralization titer (type 3/type 1)[b]		
	Preimmune	Day 20	Day 60
V_L-CSA[c] peptide	0/0	160/0	160/0
V_{L-H}-CSA peptide	0/0	20/0	160/0
Reovirus peptide	0/0	20/0	160/0

[a] Adapted from reference 132.
[b] Neutralization titer was determined by a plaque assay on murine L cells. Results represent the geometric mean for groups of four mice.
[c] CSA, Chicken serum albumin.

peptide (containing the V_H CDRII), Reo peptide (residues 317 to 332 of HA3), and V_{H-L} peptide (covalently complexed V_L and V_H peptides). Peptides were first tested for binding to Ab_1 (9BG5) either directly or by inhibition. Each of these peptides was shown to interact with 9BG5 by one of these methods (130, 131). Inhibition of 9BG5 binding to HA3 was also demonstrated for the V_L and V_{H-L} peptides. Inhibition was nearly 100% at 250 μM. The V_L peptide was further demonstrated to inhibit the binding of reovirus type 3 (75% inhibition), but not reovirus type 1, to murine L cells. From these studies, it was concluded that the V_L peptide contained the most relevant HA3 and 9BG5 binding site.

One of the most exciting prospects for an anti-idiotope-based vaccine is the use of synthetic peptides that correspond to the internal image domain of the anti-idiotope. In the most idealized case these peptides have all the advantages of anti-idiotope specificity and immunogenicity, are easily synthesized in large quantities, and are free of heterologous antibody. To test this model vaccine, mice were immunized with the peptides described above, and anti-HA3- and HA3-specific T-cell responses were measured. Mice immunized with either V_L, V_{H-L}, or Reo peptide developed anti-HA3 responses that represented 14 to 20% of the total antipeptide response (132). Direct analysis of the neutralizing activity of this antiserum is presented in Table 5. Significant inhibition of reovirus type 3 plaque formation was detected on day 20 postimmunization with V_L peptide and by day 60 with V_{H-L} and Reo peptides. There was no inhibition of reovirus type 1 plaques at any time. These results confirm that the type 3 neutralizing epitope is defined by residues 323 to 332 of HA3, which are shared by the Reo and V_L peptides. Similar results were obtained when DTH responses were analyzed. The magnitude of peptide responses was 55 to 60% of that seen with reovirus type 3.

A slightly different picture emerged in the analysis of cytolytic T cell (T_{CTL}) and T-helper (T_H) responses. In these instances, maximal reactivity was detected to either V_{H-L} or V_H peptide (132). Proliferation and interleukin-2 production by 87.92.6 or reovirus type 3-primed helper T cells was seen with V_H but not V_L peptide. These results indicate that B and T-delayed type hypersensitivity (T_{DTH}) cells recognize different domains of internal image than either T_{CTL} or T_H cells. This may relate to differences in the processing, presentation, or recognition of these different domains to and by target cells. Further studies are needed to fully resolve this issue. It is, however, clear that the anti-idiotope internal image can be related to primary sequence identity with the related epitope. How universal this phenomenon is awaits similar investigations in other systems. We are aware of one other example of immunologic mimicry by a peptide derived from the CDR of a monoclonal antireceptor antibody (115). It is equally possible that internal-image behavior can be the product of conformational mimicry that is unrelated to primary sequence homology. An obvious example of this is mimicry of nonpolypeptides, e.g., steroids and alkaloids, by antibody. These alternatives have been recently reviewed by Erlanger (24).

Development of T-cell receptor anti-idiotopes

A comprehensive study of T-cell receptor idiotype–anti-idiotype cascades for Sendai virus has been conducted by the laboratories of Ertl and Finberg. Anti-idiotypic antibodies were prepared against an IA^d-restricted, Sendai virus-specific T_H clone termed 2H3.E8 (25). Transfer of this clone had no effect on viral titers or pathogenicity in secondary hosts. Antibodies were initially screened by binding to 2H3.E8 cells. One reactive clone was identified, and subclones were named 1B4.E6 and 1B4.H6. The 1B4.E6/H6 antibody bound to 30% of all Sendai virus-reactive clones tested, but not to naïve T cells or T cells with alternative specificities (26). This antibody does not bind Sendai virus. Expression of the 1B4.E6/H6 idiotope was not restricted to either the H-2 haplotype or the IgH allotype. Anti-idiotope also bound to Sendai virus-specific CTL clones, and in agreement with the reovirus data, the same CTL clones lysed the 1B4.E6/H6 hybridoma (27).

The most probable explanations for these results are that 1B4.E6/H6 binds directly to a recurrent idiotope on the T-cell receptor of Sendai virus-reactive clones or displays an internal image of Sendai virus. In the latter instance, the response to this related epitope would have to be a dominant (approximately 30%) anti-Sendai virus response, as is seen in the HA3 response of reovirus. Other concerns are that since free virus does not bind T-cell receptor, modifications in affinity or valency would be necessary to account for direct internal-image anti-idiotope, MHC-nonrestricted binding to T-cell receptors.

The immunogenicity of 1B4.E6/H6 was evaluated by several criteria.

DTH responses were induced in mice following 1B4.E6/H6 immunization. These responses were remarkable in that T cells induced by anti-idiotope showed a lack of H-2 restriction on adoptive transfer, whereas parental 2H3.EB T cells were H-2 restricted in similar DTH response assays (29). Immunization with anti-idiotope also primed for the in vitro generation of Sendai virus-specific CTL responses (25). This included lysis of the 1B4.E6/H6 hybridoma. Once again, lysis of this hybridoma target was demonstrated by using both MHC-matched and mismatched strains.

To test for the manipulation of B-cell responses by a T-cell defined anti-idiotope, sera of mice injected intraperitoneally with crude 1B4.E6/H6 in the absence of carrier or adjuvant were examined for anti-Sendai virus reactivity. Strong IgM anti-Sendai virus responses were detected in several mouse strains by day 7 after immunization (27). Levels were equivalent to those observed with virus immunization. In each instance the level of anti-anti-idiotope accounted for only 10% of the anti-Sendai virus response. This was determined by absorption of sera with 1B4.H6-coated Sephadex beads. Serum IgG responses to anti-idiotope were weaker and were not detected until week 4. In contrast, IgG responses to immunization with Sendai virus were detected after 4 to 5 days. The anti-idiotope failed to induce antibody responses in nude mice, suggesting that these events are under T-cell control. The prophylactic effects of 1B4.E6/H6 immunization were demonstrated in that mice immunized with anti-idiotope showed increased survival and a 3-log reduction in Sendai virus titer following challenge with a lethal dose of Sendai virus. Protection was not as effective as that seen with Sendai virus preimmunization, but it was virus specific.

Taken together, these results indicate that T-cell receptor idiotope–anti-idiotope networks can be manipulated to enhance virus-specific immunity as described above for immunoglobulin idiotopes. The characteristics of the results obtained with Sendai virus indicate the recognition of a recurrent or regulatory idiotope. The most telling results are the lack of MHC restriction in T-cell responses and the small percentage of anti-anti-idiotope in the antivirus response.

Survey of investigations on virus anti-idiotypes

Our previous discussion highlights the most salient features of the most highly developed virologic systems for the study of idiotype–anti-idiotype interactions and disease. Many other laboratories have made outstanding contributions to the development of this field. Perhaps most noteworthy is the seminal work of Reagan et al. with rabies virus (93). Several new observations also offer exciting promise for future advances. All of these observations are presented for reference in outline form in Table 6.

Table 6. Summary of the Experimental use of Anti-Idiotypes as Virus Vaccines

Virus studied	Anti-idiotype form	Results of Ab_2 immunization	References
Coxsackievirus	Polyclonal	Increased macrophage chemotaxis	88
Hepatitis B virus	Polyclonal	Induction of silent clone	100
		Induction of Ab_1'	57, 61, 66
		Protective Ab_1' in chimpanzees	62
	Monoclonal	Induction of anti-HBsAg	16, 17
	Monoclonal	Ab_2 affinity and repertoire comparison	116, 117
Herpes simplex virus			
Type I	Polyclonal	Induction of DTH response	43
	Polyclonal	Induction of split tolerance and suppression of DTH	74
Type II	Polyclonal	Increased pathogenicity in mice	58
HIV-1	Polyclonal	Induced silent Ab_1'	134
	Monoclonal	Ab_1, primed or induced, Ab_2 with anti-gp120 activity	10, 11, 95, 99
		Partial neutralizing activity	21, 87
Moloney murine leukemia virus	Monoclonal	Induction of Ab_2 and Ab_3 by Ab_1	90
Mouse mammary tumor virus	Polyclonal	Induced Ab_1' and DTII	109
Newcastle disease virus	Polyclonal	Increased anti-HA response	114
Poliovirus type II	Monoclonal	Induced virus-neutralizing Ab_1'	123, 124
Pseudorabies virus	Monoclonal	Induced neutralizing Ab_1', protective Ab_1'	47
Rabies virus	Polyclonal	Induced neutralizing Ab_1'	92, 93
	Monoclonal	Induced neturalizing Ab_1'	123
Reovirus type 3	Polyclonal	Induced Ab_1', DTH and CTL responses	42, 84, 85, 105
	Monoclonal	Induced Ab_1', DTH and CTL responses	42, 104, 105
		Induced protective Ab_1'	42
	Synthetic peptides	Induced Ab_1', DTH and CTL responses	130, 132
Sendai virus	Monoclonal T idiotope	Induced Ab_1', DTH and CTL responses	25–27, 29
		Protective Ab_1'	27
	Monoclonal B idiotope	Induced Ab_1'	31
Tobacco mosaic virus	Polyclonal	Induced Ab_1'	121, 122
		Induced silent clone	36
Venezuelan equine encephalomyelitis virus	Polyclonal	Induced Ab_1'	92

APPLICATION OF ANTI-IDIOTYPES TO THE DEVELOPMENT OF AIDS VACCINES

The HIV-1 Retrovirus

This section will discuss the salient features of a recently described human pathogen, the HIV-1 retrovirus, as it relates to anti-idiotypes and the immune response. The AIDS pandemic carries profound implications for modern medical science. HIV-1 is the etiologic agent of AIDS and a spectrum of related disorders (3, 37). The clinical manifestations of this disease include immune-system failure and dementia. Immune-system destruction appears progressive and irreversible. It is believed that close to 100% of the approximately 1 million anti-HIV-positive individuals will eventually progress to the active disease state. Even more disturbing is the apparent 100% mortality rate associated with this transition (72).

HIV-1 is similar in its genomic and structural organization to other previously characterized retroviruses (3, 37, 72). It is most closely related to the lentivirus family within this group. HIV virions are composed of dimers of single-stranded RNA that are organized around products of the viral *gag* gene. These, in turn, are assembled into electron-dense cylindrical cores. The core is surrounded by a lipid bilayer as it buds across infected-cell membranes. Two viral envelope glycoproteins are displayed on the outside of the virus: gp41, which is an integral membrane protein, and gp120, which is entirely external and functions as the initial contact site for binding to the cellular receptor for virus on target cells (70). These two proteins induce the major cellular and humoral immune responses and have been the focus of vaccine development strategies (45).

HIV-HIV Receptor Interactions and HIV Infection

AIDS is characterized by both immune-system failure and neurological disorders. These clinical signs relate in part to the cellular tropism of HIV for helper T cells, astrocytes, macrophages, and histiocytic cell lineages. The selectivity of HIV binding is determined by interaction of the gp120 protein with a high-affinity binding site on the CD4 surface molecule (19, 53, 67, 71, 89). Human CD4 is a 55,000-dalton glycoprotein that participates in class II MHC-restricted activation events on T lymphocytes. Tissue surveys have demonstrated that CD4 is expressed on helper T cells, antigen-presenting cells, and rare B-cell tumor lines (79). The specific sites of interaction of gp120 and CD4 have both been determined. Site-directed mutagenesis has identified three discontinuous regions of gp120, at amino acid residues 363, 419, and 473, that participate in the CD4-binding domain (70). Similar results were obtained by peptide inhibition. These studies identified residues 397 to 439, located in the center of the area defined by mutagenesis (73).

Thus, it is likely that the viral epitope involved in CD4 binding is nonlinear and may present unique problems for immune recognition.

The binding site on CD4 is a linear epitope composed of amino acids within the amino-terminal cysteine-bonded loop (53, 71, 89). This site spans residues 30 to 60, with residues 45 to 55 as the focus. After the initial attachment of virus to CD4, other regions of the virus envelope are believed to be important in initiating fusion between the viral and cellular membranes (12, 126, 127). These secondary events precede virus entry, uncoating, and replication.

Several features of the HIV life cycle present unique problems for control of infection (75). Penetration of the virus core is followed by conversion of virus RNA into a circular DNA intermediate. This DNA is translocated to the nucleus and incorporates into the host genome. These events precede the production of virus mRNA and protein. At this point the virus may assume a dormant state, where it is isolated from the immune system. Infection of cells within the brain also presents an acute problem for immune recognition. Therefore, the challenge of AIDS vaccines is to develop an immune response of sufficiently high titer to prevent infection upon initial exposure. This approach is quite different from other vaccine strategies, which rely on boosting of responses to combat less insidious viruses.

HIV-1 and -2 are also noteworthy for an extremely high in vivo mutation rate (97). Isolates from a single individual may diverge more than 70% within regions of gp120 (35), and viruses isolated later in infection appear to be more virulent than those present initially. These observations suggest that attenuated virus may mutate to a more infectious form. Killed-virus preparations also present problems. In most instances it is difficult to ensure complete inactivation of all viruses in a preparation. The risks and consequences of infection from either killed or attenuated HIV vaccines are thus too great for general use.

Anti-Idiotypes as AIDS Vaccine Candidates

A vaccine preparation built around anti-idiotypes may have significant advantages over killed or attenuated viruses. For example, anti-idiotypes would possess no infective risk and the immune response may be targeted to a single range of stable epitopes. This is particularly important in light of the conservation of the attachment site for CD4 and the high variability of HIV-1 outside of this site. Focused antibody responses may also avoid antigenic epitopes that have been associated with an enhancing effect through increased macrophage uptake of virus.

The HIV-1 envelope proteins are composed of alternating, highly variable antigenic regions and more conserved, less antigenic regions. For this reason most patient sera demonstrate preferential type-specific over group-specific neutralization titers (45). Immunization of animals with recombinant envelope glycoproteins reveals a similar pattern of immune responsiveness. On the basis of the high variability of HIV in vivo, the type-specific neutralization response is thought to be one way in which this pathogen escapes immune destruction. Immunization with anti-idiotypic antibodies that mimic the more conserved, poorly antigenic regions of the envelope could serve to focus the immune response to more beneficial epitopes.

This strategy has had preliminary success. On the basis of the detection of a group-specific neutralization epitope at the carboxy terminus of gp41, this epitope was chosen as a candidate for anti-idiotype-based immunizations (20). Idiotypes raised in chimpanzees against a group-specific neutralizing epitope of gp41 were used as immunogens in rabbits to prepare anti-idiotypes. Anti-idiotypes were injected into mice, and sera from these animals contained antibody that reacted with the original peptide immunogen and the envelope precursor gp160 (cleaved to yield gp41 and gp120) (134). This response is uncharacteristic in that this idiotope is not normally expressed in the murine response to gp41 peptide. Thus, anti-idiotype stimulated the expression of a silent anti-gp41 clone. More expanded studies are necessary to assess the utility of this interesting approach.

The HIV-CD4 interaction presents a particularly attractive target for conformationally restricted immune responses (81). Antibodies that inhibit interactions of the virus envelope with its cellular receptor have been shown to block virus binding and infection (67, 79, 81). However, few AIDS patients develop neutralizing antibodies of this type. This may be a consequence of the nonlinearity of the gp120-binding site for CD4. This is underscored by the observation of high sequence variability in regions adjacent to the CD4-binding domain (45, 70). Conservation of the overall conformation of this region is, however, clearly required to preserve CD4 binding. As a result of this observation, several laboratories have examined the ability of anti-CD4 antibodies to induce anti-idiotypic immune responses. The Leu3a and OKT4a anti-CD4 monoclonal antibodies are of particular importance, since they may contain the internal image of the HIV-binding domain. Both antibodies bind near the amino-terminal domain of CD4 and block virus infection and gp120-induced syncytia formation.

Immune responses to Leu3a idiotypes have been studied in detail. Western immunoblot analysis shows that mice immunized with Leu3a develop responses to HIV-infected cells, HIV envelope proteins, and gp120 (11). These results were surprising, since Leu3a has poor to no reactivity to CD4 in Western analysis. The anti-Leu3a sera (Ab_2) also partially neutralized HIV infectivity. Preliminary reports from other laboratories on the success of anti-CD4 reagents as immunogens is now accumulating (87, 95,

99). The most exciting report demonstrated the induction of antibody with the ability to mediate complement-dependent lysis of HIV-infected cells (87). Whether these responses reflect internal-image or regulatory idiotopes is uncertain. Immunization of rabbits with these anti-CD4 reagents has failed to generate a neutralizing HIV response (81). However, both Leu3a and OKT4a were successful immunogens in baboons (10). A T-cell response to gp120 was also detected in these animals. Although these initial results are encouraging, specific problems remain. Most importantly, the anti-idiotypic responses displayed only limited ability to neutralize virus. This might be expected if the anti-CD4/anti-idiotype represented a poor image of the related epitope on HIV, or if responses were linked to the activation of regulatory idiotypes.

Another serious consideration of anti-CD4 vaccines is that injection of anti-CD4 may effect potent autoimmune reactivity either directly or through the formation of Ab_3. The induction of anti-CD4 antibody might also arise in the natural course of HIV infection and may contribute to reduced T-cell function. In mice, the administration of anti-CD4 antibodies results in profound nonresponsiveness to antigens present at the time of antibody administration (112, 129). Studies of AIDS patients have yielded conflicting data. One promising study, conducted by Lundin et al. (78), identified antibody that inhibited the binding of anti-CD4 to CD4 in the sera of 3 of 58 asymptomatic patients. The anti-CD4 antibodies tested all mapped to the gp120-binding site on CD4; thus, HIV entry may also be inhibited by this serum.

Experimental Approaches to HIV Biology

Alternative strategies for vaccine intervention in HIV infection could be directed at other steps in the HIV life cycle. It is now apparent that HIV entry into cells is a complex process. CD4 expression does not alone confer susceptibility to infection. Murine cells that express human CD4 are not infected, despite binding HIV-1 in a manner indistinguishable from their human counterparts (79, 127, 128). The barrier to this infection does not reside with the cellular machinery of murine cells. Direct electroporation of HIV-1 proviral clones into the same murine cells leads to a normal infectious cycle (127). CD4-independent infection of human cells has also been reported (12). These results imply that surface molecules ancillary to CD4 are important in mediating HIV entry.

One approach used to identify these ancillary components was to screen patient sera for antibody that altered virus-cell interactions. Some AIDS patient sera contained low levels of antibodies that inhibited syncytia formation (128). These sera also showed reaction with viral envelope determinants other than those involved in gp120-CD4 binding. One patient,

Table 7. Molecular Constraints on HIV-1 Infectivity

Cell line	Source	Antibody binding[a]		HIV-1 infection[b]
		Anti-CD4	Anti-H156	
Sup T1	Human	+	+	+
CEM	Human	+	+	+
H9	Human	+	+	+
U937a	Human	+	−	−
HSB	Human	−	+	−
HSB/CD4	Human	+	+	+
HeLa	Human	−	+	−
HeLa/CD4	Human	+	+	+
L	Murine	−	−	−
L/CD4	Murine	+	−	−
NIH 3T3	Murine	−	−	−
NIH 3T3/CD4	Murine	+	−	−

[a] Determined by fluorescence-activated cell sorter analysis.
[b] Determined by reverse transcriptase assay after coculture with SupT1 cells.

termed H156, had high levels of such activity. To characterize the nature of this reactivity and its target sites, anti-idiotypic antibodies were constructed (126). The rationale for this internal-image mimicry is presented more fully in Idiotypic Reagents as Tools for the Dissection of Virus Biology (see below). Purified anti-idiotype displayed specific binding to H156 and pooled AIDS immunoglobulin, indicative of recurrent idiotopes in the anti-HIV response. Binding analyses indicated that this anti-idiotype (anti-H156) recognized a molecule present on human but not murine cells. Indeed, the coexpression of this structure with CD4 was correlated with the species tropism reported for HIV-1. These results are presented in summary form in Table 7. Studies are currently under way to determine the structure and function of this molecule and to assess the efficacy of antibodies to this structure in the inhibition of HIV infection.

IDIOTYPIC REAGENTS AS TOOLS FOR THE DISSECTION OF VIRUS BIOLOGY

The Basis of Idiotypic Mimicry

The preceding sections outlined the structural and biological characteristics of anti-idiotypic antibodies. They illustrated that one characteristic of internal-image antibody is epitope mimicry. If this epitope lies at a relevant site of binding or attachment of the antigen to another molecule, anti-idiotope may mimic this interaction. Sege and Peterson were the first to demonstrate idiotypic mimicry of biological receptors (101). They showed that anti-idiotypic antibody to retinol-binding protein mimicked retinol-binding protein binding to human prealbumin and inhibited retinol-binding protein-mediated uptake of retinol by intestinal epithelial cells (102). They were also the first to demonstrate anti-idiotypic mimicry of ligand signal delivery. Anti-idiotypic antibodies prepared to insulin inhibited insulin binding and stimulated the insulin-dependent uptake of α-aminoisobutyric acid in thymocytes. There are now numerous examples of idiotypic mimicry of ligand and receptor function (reviewed in reference 39). Several laboratories have adopted this approach for the study of virus-virus receptor interactions. The most thorough approaches have been with reovirus and Semliki Forest virus.

Experimental Applications of Anti-Idiotypic Probes

Isolation and function of reovirus receptors

Anti-idiotypic antibodies to reovirus HA3 have been used as probes of the biochemistry and cell biology of the mammalian reovirus receptor. The binding of reovirus to target cells is governed by the specificity of the minor outer capsid protein S1. This protein also serves as the viral HA. Results presented in Examination of the Structural Basis of Internal Image (see above) indicate that in addition to tissue tropism, the reovirus HA determines the serotype specificity of antibody, CTL, DTH, and helper T immune responses. The production of polyclonal and monoclonal anti-idiotope to reovirus HA3 was also described previously (39). Internal image was confirmed by several immunologic criteria and ultimately by primary sequence similarity (7, 42, 104, 105). One of the earliest observations was that the pattern of anti-idiotope binding mimicked the tissue tropism of reovirus type 3 to both central nervous system and lymphoid-cell populations (55, 84). More direct evidence that anti-idiotope recognized the type 3 receptor was obtained from binding inhibition studies. Anti-idiotope inhibited virus binding to target cells by approximately 90% (38, 39, 55) (Table 8). The ability of anti-idiotope to inhibit infectivity was reported by Dichter et al. (22). A 60-min preincubation of explanted neurons with anti-idiotope reduced subsequent reovirus infection in these cultures by 75%.

On the basis of these studies, anti-idiotope was used for the isolation and biochemical characterization of reovirus type 3 receptors. Reovirus receptors were isolated from a panel of cell lines with polyclonal and monoclonal anti-idiotype (14, 38). The cell surface proteins were first labeled with ^{125}I, solubilized in detergent, and then immunoprecipitated with anti-idiotype. A representative example of the sodium dodecyl sulfate-polyacrylamide gel electrophoresis analysis of such an immunoprecipitation is shown

Table 8. Anti-Idiotope Inhibition of Reovirus Binding[a]

Cell line	Mean binding ± SEM (cpm bound)		% Inhibition of virus binding
	Untreated cells	Treated with anti-idiotope	
Idiotope positive			
R1.1	3,319 ± 530	317 ± 416	90.4
R1.E	3,355 ± 348	365 ± 183	89.1
Idiotope negative			
EL.4	7,355 ± 1,832	6,022 ± 77	18.1
P815	15,380 ± 549	13,356 ± 622	13.2

[a] From reference 39. Binding studies were conducted at 4°C by the addition of 25,000 cpm of ^{35}S-labeled reovirus type 3 particles to 2×10^6 cells from 1 h. The effect of anti-idiotope was determined by preincubation of cells with purified antibody for 30 min prior to virus addition. Untreated controls were without antibody addition.

in Fig. 2. A single band with an apparent molecular mass of 67,000 daltons was uniformly observed in all receptor-positive cell types tested. These observations were strengthened by Western blot analysis of cell membranes run on sodium dodecyl sulfate-polyacrylamide gel electrophoresis (14, 38). Reovirus and polyclonal and monoclonal anti-idiotope all bound to a similar 67,000-dalton band that was not observed in any membrane or antibody controls. Monoclonal anti-idiotope has also been used to measure the number of receptors on target cells. Equivalent numbers of receptors were de-

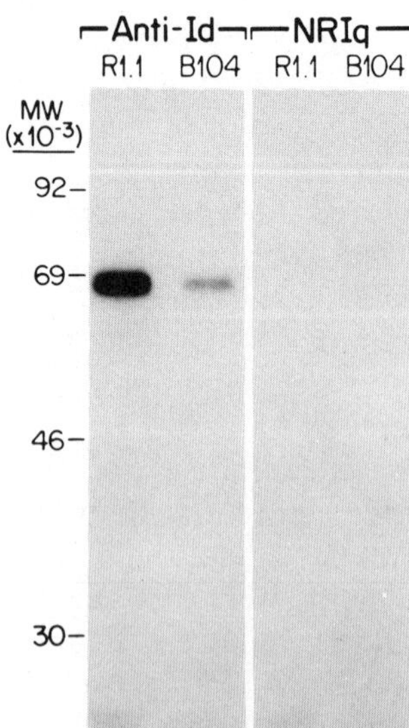

Figure 2. Immunoprecipitation of reovirus receptor with anti-idiotype. Sodium dodecyl sulfate-polyacrylamide gel analysis under reducing conditions of ^{125}I-labeled cell membranes immunoprecipitated with either rabbit anti-idiotype (lanes 1 and 2) or normal rabbit immunoglobulin (lanes 3 and 4). Lanes 1 and 3 show immunoprecipitates from labeled R1.1 rat thymoma cells, and lanes 2 and 4 show immunoprecipitates from labeled B104 rat neuroblastoma cells. (Adapted from reference 14.)

tected on several cell lines by using either antibody or virus probes (14, 15, 38).

Data from a number of systems indicate that many viruses utilize integral membrane proteins that provide essential cellular functions as viral attachment sites. For example, the membrane receptor for Epstein-Barr virus has been shown to bind to the C3d receptor CR2 on B lymphocytes (34), and the CD4 molecule serves as the attachment site for HIV on T lymphocytes (19). Recent experiments, conducted with anti-idiotypic probes, suggest that a structural homolog of the β-adrenergic receptor serves as the reovirus type 3 attachment site.

Initial investigations indicated that the tissue distribution and molecular weight of the mammalian β-adrenergic receptor matched those of reovirus receptors. More detailed studies have shown that anti-idiotope specifically immunoprecipitated purified adrenergic receptor (15); the two-dimensional gel patterns and partial tryptic digests of purified adrenergic and reovirus receptors were indistinguishable; and anti-idiotope coprecipitated labeled β-adrenergic ligands bound on the cell surface. Recent studies conducted with purified reovirus receptors demonstrated that β-ligand binding to reovirus receptor is specific but of reduced affinity when compared with binding to purified adrenergic receptors (G. Gaulton, unpublished data). Analysis of viable cells indicated that reovirus and β-ligands do not show competition for binding and that reovirus binding does not stimulate classical β-adrenergic signaling pathways (41). These results indicate that reovirus does not bind to functional β-adrenergic receptors, but probably does bind to a related member of the adrenergic receptor family.

One early consequence of the attachment of reovirus type 3 to cells is a dramatic inhibition of host cell DNA synthesis. Inhibition is first detected at 8 to 10 h postinfection, and synthesis is only 15 to 25% of uninfected levels by 24 h (23, 103). Several observations hinted that the mechanism of this effect was linked to a direct perturbation of the reovirus receptor and not to subsequent reovirus gene activity or host alteration. The most important of these was that inhibition was absolutely linked to the presence of the HA3 molecule (103). Anti-idiotope that mimicked HA3 binding offered the ideal probe to analyze this interaction in the absence of secondary virus effects. These studies clearly demonstrated that receptor binding and aggregation was the trigger for DNA inhibition (41).

The addition of purified anti-idiotope to a panel of rat neuroblastoma lines caused an inhibition of DNA synthesis within 12 h after binding (41). The kinetics of this inhibition were identical to those seen with reovirus type 3. A suggestion that inhibition was dependent on receptor aggregation was made from results that 50% inhibition was observed at 70% receptor occupancy. This was confirmed by showing that inhibition was not seen with monovalent Fab fragments and that it was maximized by addition of anti-immunoglobulin to fragments or Fab coupled to beads. Reversible-binding analysis indicated that the signals delivered by anti-idiotope occurred within 30 to 60 minutes after initial binding to receptor. Thus, the use of this anti-idiotypic, antireceptor probe has helped to define both the molecular biology of reovirus and reovirus pathogenicity.

Anti-idiotypic probes of Semliki Forest virus nucleocapsid interactions

Internal-image anti-idiotopes have also been used to explore the interaction of Semliki Forest virus capsid-spike components during virus assembly. The budding of enveloped viruses such as Semliki Forest virus is characterized by selective packaging and interaction of viral components. Semliki Forest virus packaging begins by enclosure of the genomic RNA into the nucleocapsid (125). The capsid then attaches to the plasma membrane and associates with the spike glycoprotein complex to form the nascent virion. A specific interaction between the nucleocapsid and the cytoplasmic domain of the spike glycoprotein complex has been proposed as a mechanism for this directed and selective assembly (108).

The nature of the capsid-spike interaction was defined by using an idiotypic cascade. Antibodies to the cytoplasmic domain of the E2 spike glycoprotein were prepared by using a synthetic peptide (anti-E2). This peptide was also used for in vitro immunization. Supernatants of in vitro reactions that were positive for anti-E2 activity were used for a second round of in vitro immunization to prepare the anti-idiotope (anti-anti-E2). Monoclonal antibodies were constructed from this second round, and one clone, designated F13, was shown to bear the internal image of E2 peptide by its ability to bind and immunoprecipitate the nucleocapsid protein. Anti-anti-idiotopes (Ab$_3$) were next prepared by in vivo immunization and fusion. One monoclonal antibody, designated 3G10, was again selected as internal image by its capacity to both compete with antigen for binding to F13 and by binding to the E2 peptide. The demonstration that monoclonal antibody F13 bound nucleocapsid and that monoclonal antibody 3G10 bound the cytoplasmic E2 peptide confirms the interaction of nucleocapsid with this cytoplasmic domain of E2.

Survey of applications of idiotypic mimicry in virology

Several laboratories have now used the approach of internal image, described above for reovirus and Semliki Forest virus, to develop probes for the dissection of virologic processes. The majority of these studies have used anti-idiotopes to either identify, isolate, or define the binding parameters of virus receptors. A complete listing of these applications is presented in Table 9.

Table 9. Summary of the Experimental Use of Idiotypic Mimicry in Virus Systems

Virus	Anti-idiotype form	Biological studies	References
Epstein-Barr virus	Polyclonal	Distinction of EBV and C3d binding sites on gp140	2
Hepatitis B virus	Monoclonal	Receptor for polyHSA defined on HBsAg	17
HIV-1	Polyclonal	Identified ancillary component of HIV entry	126–128
Leukemogenic retrovirus	Monoclonal	Isolation of virus receptor	1
Polyomavirus	Monoclonal	Identification of virus receptor; inhibition of virus binding/infectivity	80
Reovirus type 3	Monoclonal	Isolation and distribution of virus receptor	14, 38
		Homology to adrenergic receptors	15
			22
		Inhibition of virus binding/infectivity	41
		Mimicry of receptor signaling	
Semliki Forest virus	Monoclonal	Definition of capsid-spike binding and assembly	125

CONCLUSIONS AND PERSPECTIVES

The idiotypic network plays an important role in the regulation of immune responses to virus. This cascade has also been experimentally manipulated to produce effective virus vaccines and experimental probes of virus receptors. Data reviewed in the opening sections of this chapter illustrate that antibody responses to virus often display an idiotypic bias. This may be a reflection of the preferential use of particular V-region genes or the expansion and somatic mutation of B-cell clones. In either case, this idiotypic bias contributes a strong regulatory component through the interplay of so-called regulatory idiotopes on clones of both similar and seemingly unrelated specificities. Despite these observations, the exact role of idiotopes in the natural course of virus diseases remains obscure. The most promising studies are those on patients with chronic Epstein-Barr virus, in whom high anti-idiotope levels are associated with a decline in anti-Epstein-Barr virus responses that may help establish permissiveness for Epstein-Barr virus.

There is no question that the idiotypic network can be manipulated to prime, bolster, or fully induce potent B- and T-cell immunity to virus. Anti-idiotope vaccines offer considerable advantages over conventional approaches in specificity, safety, and cost effectiveness. Work with hepatitis B virus demonstrated the extension of this approach to primate vaccines. Important advances in the molecular nature of internal image by workers in the reovirus system now suggest that it will be possible to define synthetic peptide vaccines based on internal image. This will help to minimize the risk associated with injection of heterologous immunoglobulin. The future in this field seems unlimited and particularly relevant to development of an AIDS vaccine.

Internal-image anti-idiotopes have also proven to be powerful biochemical tools. There are now several established applications of internal image for identification and isolation of virus receptors. The only constraints on this approach are that some molecular interactions may have composite or conformationally severe binding sites that may not be potentially reflected in the immunoglobulin architecture. These data also provide a clear illustration of one means of autoimmunity. That is, one potential consequence of idiotypic mimicry of epitopes lying at or near virus-binding sites is the production of auto-anti-idiotypes that bind to the cellular attachment molecules of virus, thereby initiating pathogenesis. Several autoimmune diseases are related to the presence of antireceptor antibodies. These include Graves' disease and anti-thyroid stimulating hormone receptor antibody (30), myasthenia gravis and anti-acetylcholine receptor antibody (76), and certain forms of insulin-resistant diabetes mellitus and anti-insulin receptor antibody (96). We note that none of these diseases has been directly linked to a viral etiology.

We have previously proposed a model of autoimmunity that incorporates these observations (39). In the inductive phase of disease, the virus initiates an idiotypic cascade that results in the production of anti-idiotypic, antireceptor antibodies. In the expansion phase of disease these antireceptor antibodies, largely IgM, bind and destroy cellular targets through engagement of host effector responses such as complement fixation. In the regulatory phase of disease newly exposed secondary epitopes on receptors are processed and high-affinity responders are generated. Finally, in the pathogenic phase the presence of multiple, high-affinity antireceptor antibodies leads to clinical symptoms of disease. This model remains a viable and experimentally testable approach to understanding virally associated autoimmune syndromes.

Acknowledgments. We are most grateful to the many colleagues who participated in the work on the reovirus system; they include Mark Greene, Man-Surg Co, Arlene Sharpe, Bernie Fields, Bill Williams, and Jerry Nepom. We also thank Joanne Pratt for critical review of the manuscript.

G.N.G. is a Harry Weaver Neuroscience Scholar of the National Multiple Sclerosis Society.

REFERENCES

1. **Ardman, B., R. H. Khiroya, and R. S. Schwartz.** 1985. Recognition of a leukemia-related antigen by an antiidiotypic antiserum to an anti-gp70 monoclonal antibody. *J. Exp. Med.* **161**:669–686.

2. **Barel, M., A. Fiandino, A. X. Delcayre, F. Lyamani, and R. Frade.** 1988. Monoclonal and anti-idiotypic anti-EBV/C3d receptor antibodies detect two binding sites, one for EBV and one for C3d on glycoprotein 140, the EBV/C3dR, expressed on human B lymphocytes. *J. Immunol.* **141**:1590–1595.

3. **Barré-Sinoussi, F., J. C. Chermann, F. Rey, M. T. Nugeyre, S. Chamaret, J. Gruest, C. Dauguet, C. Axler-Blin, F. BrunVézinet, W. Rozenbaum, and L. Montagnier.** 1983. Isolation of a T lymphotrophic retrovirus from a patient at risk for acquired immune deficiency syndrome (AIDS). *Science* **220**:868–871.

4. **Bona, C., E. Heber-Katz, and W. E. Paul.** 1981. Idiotype-anti-idiotype regulation. I. Immunization with a levan-binding myeloma protein leads to the appearance of autoanti-(anti-Id) antibodies and to activation of silent clones. *J. Exp. Med.* **153**:951–967.

5. **Bona, C., and W. E. Paul.** 1979. Cellular basis of regulation of expression of idiotype. I. T-suppressor cells specific for MOPC 460 idiotype regulate the expression of cells secreting anti-TNP antibodies bearing 460 idiotype. *J. Exp. Med.* **149**:592–600.

6. **Bona, C. A.** 1984. Regulatory idiotypes, p. 29–43. *In* H. Köhler, J. Urbain, and P.-A. Cazenave (ed.), *Idiotypy in Biology and Medicine.* Academic Press, Inc., New York.

7. **Bruck, C., M. S. Co, M. Slaoui, G. N. Gaulton, T. Smith, B. N. Fields, J. I. Mullins, and M. I. Greene.** 1986. Nucleic acid sequence of an internal image-bearing monoclonal anti-idiotype and its comparison to the sequence of the external antigen. *Proc. Natl. Acad. Sci. USA* **83**:6578–6582.

8. **Caton, A. J., G. G. Brownlee, L. M. Staudt, and W. Gerhard.** 1986. Structural and functional implications of a restricted antibody response to a defined antigenic region on the influenza virus hemagglutinin. *EMBO J.* **5**:1577–1587.

9. **Cazenave, P.-A.** 1977. Idiotypic-anti-idiotypic regulation of antibody synthesis in rabbits. *Proc. Natl. Acad. Sci. USA* **74**:5122–5125.

10. **Chanh, T. C., B. E. Alderette, E.-M. Zhou, and R. C. Kennedy.** 1987. Anti-idiotypic antibodies against OKT4a bind to human immunodeficiency virus. *Fed. Proc.* **46**:1352. (Abstract.)

11. **Chanh, T. C., G. R. Dreesman, and R. C. Kennedy.** 1987. Monoclonal anti-idiotypic antibody mimics the CD4 receptor and binds human immunodeficiency virus. *Proc. Natl. Acad. Sci. USA* **84**:3891–3895.

12. **Clapham, P. R., J. N. Weber, D. Whitby, D. McIntosh, A. G. Dalgleish, P. J. Maddon, L. C. Deen, R. W. Sweet, and R. A. Weiss.** 1989. Soluble CD4 blocks HIV-1 infection of t cells and macrophages but not fibroblast cell lines. *Nature* (London) **337**:368–372.

13. **Clarke, S. H., K. Huppi, D. Ruezinsky, L. Staudt, W. Gerhard, and M. Weigert.** 1985. Inter- and intraclonal diversity in the antibody response to influenza hemagglutinin. *J. Exp. Med.* **161**:687–704.

14. **Co, M. S., G. N. Gaulton, B. N. Fields, and M. I. Greene.** 1985. Isolation and biochemical characterization of the mammalian reovirus type 3 cell-surface receptor. *Proc. Natl. Acad. Sci. USA* **82**:1494–1498.

15. **Co, M. S., G. N. Gaulton, A. Tominaga, C. J. Homcy, B. N. Fields, and M. I. Greene.** 1985. Structural similarities between the mammalian B-adrenergic and reovirus type 3 receptors. *Proc. Natl. Acad. Sci. USA* **82**:5315–5318.

16. **Colucci, G., Y. Beazer, and S. D. Waksal.** 1987. Interactions between hepatitis B virus and polymeric human serum albumin. II. Development of syngeneic monoclonal anti-anti-idiotypes which mimic hepatitis B surface antigen in the induction of immune responsiveness. *Eur. J. Immunol.* **17**:371–374.

17. **Colucci, G., and S. D. Waksal.** 1987. Interactions between hepatitis B virus and polymeric human albumin. I. Production of monoclonal anti-idiotypes (anti-anti-polymeric human albumin) which recognize hepatitis B virus surface antigen. *Eur. J. Immunol.* **17**:365–370.

18. **Cosenza, H., and H. Kohler.** 1972. Specific suppression of the antibody response by antibodies to receptors. *Proc. Natl. Acad. Sci. USA* **69**:2701–2705.

19. **Dalgleish, A. G., P. C. L. Beverley, P. R. Clapham, D. H. Crawford, M. F. Greaves, and R. A. Weiss.** 1984. The CD4 (T4) antigen is an essential component of the receptor for the AIDS retrovirus. *Nature* (London) **312**:763–767.

20. **Dalgleish, A. G., T. C. Chanh, R. C. Kennedy, P. Kanda, P. R. Clapham, and R. A. Weiss.** 1988. Neutralization of diverse HIV-1 strains by monoclonal antibodies raised against a gp41 synthetic peptide. *Virology* **165**:209–215.

21. **Dalgleish, A. G., B. J. Thomson, T. C. Chanh, M. Malkovsky, and R. C. Kennedy.** 1987. Neutralization of HIV isolates by anti-idiotypic antibodies which mimic the T4 (CD4) epitope: a potential AIDS vaccine. *Lancet* **ii**:1047–1050.

22. **Dichter, M. A., H. L. Weiner, B. N. Fields, G. Mitchell, J. Noseworthy, G. N. Gaulton, and M. I. Greene.** 1986. Antiidiotypic antibody to reovirus binds to neurons and protects from viral infection. *Ann. Neurol.* **19**:555–558.

23. **Ensminger, W. D., and I. Tamm.** 1969. The step in cellular DNA synthesis blocked by reovirus infections. *Virology* **39**:935–938.

24. **Erlanger, B. F.** 1989. Auto-anti-idiotypy, autoimmunity and some thoughts on the structure of internal images. *Int. Rev. Immunol.* **5**:131–137.

25. **Ertl, H. C. J., and R. W. Finberg.** 1984. Sendai virus-specific T-cell clones: Induction of cytolytic T cells by an anti-idiotypic antibody directed against a helper T-cell clone. *Proc. Natl. Acad. Sci. USA* **81**:2850–2854.

26. **Ertl, H. C. J., and R. W. Finberg.** 1984. Characteristics and functions of Sendai virus-specific T-cell clones. *J. Virol.* **50**:425–431.

27. **Ertl, H. C. J., and R. W. Finberg.** 1986. Use of T cell-specific anti-idiotypes to immunize against viral infections. *Immunol. Rev.* **90**:129–155.

28. **Ertl, H. C. J., M. I. Greene, J. H. Noseworthy, B. N. Fields, J. T. Nepom, D. R. Spriggs, and R. W. Finberg.** 1982. Identification of idiotypic receptors on reovirus-specific cytolytic T cells. *Proc. Natl. Acad. Sci. USA* **79**:7479–7483.

29. **Ertl, H. C. J., E. Homans, S. Tournas, and R. W. Finberg.** 1984. Sendai virus-specific T cell clones. V. Induction of a virus-specific response by antiidiotypic antibodies directed against a T helper cell clone. *J. Exp. Med.* **159**:1778–1783.

30. **Farid, N. R., and T. C. Lo.** 1985. Antiidiotypic antibodies as probes for receptor structure and function. *Endocr. Rev.* **6**:1–23.

31. **Fenner, M., K. Siegmann, and H. Binz.** 1986. Monoclonal antibodies specific for Sendai virus. II. Production of monoclonal anti-idiotypic antibodies. *Scand. J. Immunol.* **24**:341–349.

32. **Fields, B. N., and M. I. Greene.** 1982. Genetic and molecular mechanisms of viral pathogenesis: implications for prevention and treatment. *Nature* (London) **300**:19–23.

33. **Finberg, R., and B. Benacerraf.** 1981. Induction, control, and consequences of virus specific cytotoxic T cells. *Immunol. Rev.* **58**:157–180.

34. **Fingeroth, J. D., J. J. Weiss, T. F. Tedder, J. L. Strominger, P. A. Biro, and D. T. Fearon.** 1984. Epstein-Barr virus receptor on human B lymphocytes is the C3d receptor CR2. *Proc. Natl. Acad. Sci. USA* **81**:4510–4514.

35. **Fisher, A. G., B. Ensoli, D. Looney, A. Rose, R. C. Gallo, M. S. Saag, G. M. Shaw, B. H. Hahn, and F. Wong-Staal.** 1988. Biologically diverse molecular variants within a single HIV-1 isolate. *Nature* (London) **334**:444–447.

36. **Francotte, M., and J. Urbain.** 1984. Induction of anti-tobacco mosaic virus antibodies in mice by rabbit antiidiotypic antibodies. *J. Exp. Med.* **160**:1485–1494.

37. **Gallo, R. C., P. S. Sarin, E. P. Gelmann, M. Robert-Guroff, E. Richardson, V. S. Kallyanaraman, D. Mann, G. D. Sidhu, R. E. Stahl, S. Zolla-Pazner, J. Leibowitch, and M. Popovic.** 1983. Frequent detection and isolation of cytopathic retroviruses from patients with AIDS and at risk for AIDS. *Science* **220**:865–867.

38. **Gaulton, G. N., M. S. Co, and M. I. Greene.** 1985. Anti-idiotypic antibody identifies the cellular receptor of reovirus type 3. *J. Cell. Biochem.* **28**:69–78.

39. **Gaulton, G. N., and M. I. Greene.** 1986. Idiotypic mimicry of biological receptors. *Annu. Rev. Immunol.* **4**:253–280.

40. **Gaulton, G. N., and M. I. Greene.** 1986. Anti-idiotypic antibodies of reovirus as biochemical and immunological mimics. *Int. Rev. Immunol.* **1**:79–90.

41. **Gaulton, G. N., and M. I. Greene.** 1989. Inhibition of cellular DNA synthesis by reovirus occurs through a receptor-linked signaling pathway that is mimicked by antiidiotypic, antireceptor antibody. *J. Exp. Med.* **169**:197–211.

42. **Gaulton, G. N., A. H. Sharpe, D. W. Chang, B. N. Fields, and M. I. Greene.** 1986. Syngeneic monoclonal internal image anti-idiotopes as prophylactic vaccines. *J. Immunol.* **137**:2930–2936.

43. **Gell, P. G. H., and P. A. H. Moss.** 1985. Production of cell-mediated immune response to herpes simplex virus by immunization with anti-idiotypic heteroantisera. *J. Gen. Virol.* **66**:1801–1804.

44. **Gerhard, W., J. Yewdell, M. E. Frankel, and R. Webster.** 1981. Antigenic structure of influenza virus hemagglutinin defined by hybridoma antibodies. *Nature* (London) **290**:713–717.

45. **Goulsmit, J.** 1988. Immunodominant B-cell epitopes of the HIV-1 envelope recognized by infected and immunized hosts. *AIDS* **2**(Suppl.):41–45.

46. **Green, D. R., P. M. Flood, and R. K. Gershon.** 1983. Immunoregulatory T-cell pathways. *Annu. Rev. Immunol.* **1**:439–498.

47. **Gurish, M. F., T. Ben-Porat, and A. Nisonoff.** 1988. Induction of antibodies to pseudorabies virus by immunization with antiidiotypic antibodies. *Ann. Inst. Pasteur Immunol.* **139**:677–687.

48. **Hiernaux, J. R.** 1988. Idiotypic vaccines and infectious disease. *Infect. Immun.* **56**:1407–1413.

49. **Huber, R., J. Deisenhofer, P. M. Colman, M. Matshushima, and W. Palm.** 1976. Crystallographic structure of an IgG molecule and an Fc fragment. *Nature* (London) **264**:415–420.

50. **Ibarra, M. Z., I. Mora, J. A. Quiroga, J. Bartolome, F. La Banda, J. C. Porres, and V. Carreno.** 1988. IgG and IgM autoantiidiotype antibodies against antibody to HBsAg in chronic hepatitis B. *Hepatology* **8**:775–780.

51. **Imboden, J. B., A. Weiss, and J. D. Stobo.** 1985. The antigen receptor on a human T cell line initiates activation by increasing cytoplasmic free calcium. *J. Immunol.* **134**:663–665.

52. **Ionescu-Matiu, I., R. C. Kennedy, J. T. Sparrow, A. R. Culwell, Y. Sanchez, J. L. Melnick, and G. R. Dreesman.** 1983. Epitopes associated with a synthetic hepatitis B surface antigen peptide. *J. Immunol.* **130**:1947–1952.

53. **Jameson, B. A., P. E. Roa, L. I. Kong, B. H. Hahn, G. M. Shaw, L. E. Hood, and S. B. H. Kent.** 1988. Location and chemical synthesis of a binding site for HIV-1 on the CD4 protein. *Science* **240**:1335–1339.

54. **Jerne, N. K.** 1974. Towards a network theory of the immune system. *Ann. Immunol.* (Paris) *Sect. C* **125**:373–389.

55. **Kauffman, R. S., J. H. Noseworthy, J. T. Nepom, R. Finberg, B. N. Fields, and M. I. Greene.** 1983. Cell receptors for mammalian reovirus. II. Monoclonal anti-idiotypic antibody blocks viral binding to cells. *J. Immunol.* **131**:2539–2541.

56. **Kees, U. R.** 1981. Idiotypes on major histocompatibility complex-restricted virus-immune cytotoxic T lymphocytes. *J. Exp. Med.* **153:**1562–1573.

57. **Kennedy, R. C.** 1985. Idiotype networks in hepatitis B virus infections. *Curr. Top. Microbiol. Immunol.* **119:**1–13.

58. **Kennedy, R. C., K. Adler-Storthz, J. W. Burns, Sr., R. D. Henkel, and G. R. Dreesman.** 1984. Antiidiotype modulation of herpes simplex virus infection leading to increased pathogenicity. *J. Virol.* **50:**951–953.

59. **Kennedy, R. C., K. Adler-Storthz, R. D. Henkel, Y. Sanchez, J. L. Melnick, and G. R. Dreesman.** 1983. Immune response to hepatitis B surface antigen: enhancement by prior injection of antibodies to the idiotype. *Science* **221:**853–855.

60. **Kennedy, R. C., and G. R. Dreesman.** 1983. Common idiotypic determinant associated with human antibodies to hepatitis B surface antigen. *J. Immunol.* **130:**385–389.

61. **Kennedy, R. C., and G. R. Dreesman.** 1984. Enhancement of the immune response to hepatitic B surface antigen. In vivo administration of antiidiotype induces anti-HBs that express a similar idiotype. *J. Exp. Med.* **159:**655–665.

62. **Kennedy, R. C., J. W. Eichberg, R. E. Lanford, and G. R. Dreesman.** 1986. Anti-idiotypic antibody vaccine for type B viral hepatitis in chimpanzees. *Science* **232:**220–223.

63. **Kennedy, R. C., I. Ionescu-Matiu, Y. Sanchez, and G. R. Dreesman.** 1983. Detection of interspecies idiotypic cross-reactions associated with antibodies to hepatitis B surface antigen. *Eur. J. Immunol.* **13:**232–235.

64. **Kennedy, R. C., J. L. Melnick, and G. R. Dreesman.** 1984. Antibody to hepatitic B virus induced by injecting antibodies to the idiotype. *Science* **223:**930–931.

65. **Kennedy, R. C., Y. Sanchez, I. Ionescu-Matiu, J. L. Melnick, and G. R. Dreesman.** 1982. A common human anti-hepatitis B surface antigen idiotype is associated with the group a conformation-dependent antigenic determinant. *Virology* **122:**219–221.

66. **Kennedy, R. C., J. T. Sparrow, Y. Sanchez, J. L. Melnick, and G. R. Dreesman.** 1984. Enhancement of viral hepatitis B antibody (anti-HBs) response to a synthetic cyclic peptide by priming with anti-idiotype antibodies. *Virology* **136:**247–252.

67. **Klatzmann, D., E. Champagne, S. Chamaret, J. Gruest, D. Guetard, T. Hercend, J.-C. Gluckman, and L. Montagnier.** 1984. T lymphocyte T-4 behaves as the receptor for the human retrovirus LAV. *Nature* (London) **312:**767–768.

68. **Kluskens, L., and H. Kohler.** 1974. Regulation of immune response by autogenous antibody against receptor. *Proc. Natl. Acad. Sci. USA* **71:**5083–5087.

69. **Koff, W. C., and D. F. Hoth.** 1988. Development and testing of AIDS vaccines. *Science* **241:**426–432.

70. **Kowalski, M., J. Potz, L. Basiripour, T. Dorfman, W. Chun Goh, E. Terwilliger, A. Dayton, C. Rosen, W. Haseltine, and J. Sodrofsky.** 1987. Functional regions of the envelope glycoprotein of human immunodeficiency virus type 1. *Science* **233:**209–212.

71. **Landau, N. R., M. Warton, and D. R. Littman.** 1988. The envelope glycoprotein of the human immunodeficiency virus binds to the immunoglobulin-like domain of CD4. *Nature* (London) **334:**159–162.

72. **Lane, H., and A. Fauci.** 1985. Infectivity and pathogenesis of AIDS. *Annu. Rev. Immunol.* **3:**477–500.

73. **Lasky, L. A., G. Nakamura, D. H. Smith, C. Fennie, C. Shimasaki, E. Patzer, P. Berman, T. Gregory, and D. J. Capon.** 1987. Delineation of a region of the human immunodeficiency virus type I gp120 glycoprotein critical for interaction with the CD4 receptor. *Cell* **50:**975–985.

74. **Lathey, J. L., S. Martin, and B. T. Rouse.** 1987. Suppression of delayed type hypersensitivity to herpes simplex virus type 1 following immunization with anti-idiotypic antibody: an example of split tolerance. *J. Gen. Virol.* **68:**1093–1102.

75. **Lawrence, J. A.** 1988. Vaccines and immunology, an overview. *AIDS* **2**(Suppl.):91–94.

76. **Lindstrom, J.** 1985. Immunobiology of myasthenia gravis, experimental autoimmune myasthenia gravis, and Lambert-Eaton syndrome. *Annu. Rev. Immunol.* **3**:109–131.

77. **Liu, Y. N., C. A. Bona, and J. L. Schulman.** 1981. Idiotypy of clonal responses to influenza virus hemagglutinin. *J. Exp. Med.* **154**:1525–1538.

78. **Lundin, K., A. Karlsson, A. Nygren, A. Lofstrom, D. Gigliotti, and H. Wigzell.** 1988. Certain human gp120-HIV antibodies react with anti-CD4 antibodies. *Scand. J. Immunol.* **27**:113–117.

79. **Maddon, R. J., A. G. Dalgleish, J. S. McDougal, P. R. Clapham, R. A. Weiss, and R. Axel.** 1985. The T4 gene encodes the AIDS virus receptor and is expressed in the immune system and the brain. *Cell* **47**:333–348.

80. **Marriott, S. J., D. J. Roeder, and R. A. Consigli.** 1987. Anti-idiotypic antibodies to a polyomavirus monoclonal antibody recognize cell surface components of mouse kidney cells and prevent polyomavirus infection. *J. Virol.* **61**:2747–2753.

81. **McDougal, J. S., J. K. A. Nicholson, G. D. Cross, S. P. Cort, M. S. Kennedy, and A. C. Mawle.** 1986. Binding of the human retrovirus HTLV-III/LAV/ARV/HIV to the CD4 (T4) molecule: conformation dependence, epitope mapping, antibody inhibition and potential for idiotypic mimicry. *J. Immunol.* **137**:2937–2944.

82. **McKean, D., K. Huppi, M. Bell, L. Staudt, W. Gerhard, and M. Weigert.** 1984. Generation of antibody diversity in the immune response of BALB/c mice to influenza virus hemagglutinin. *Proc. Natl. Acad. Sci. USA* **81**:3180–3184.

83. **Moran, T., Y. N. C. Liu, J. L. Schilman, and C. A. Bona.** 1984. Shared idiotopes among monoclonal antibodies specific for A/PR/8/34 (H1N1) and X-31 (H3N2) influenza viruses. *Proc. Natl. Acad. Sci. USA* **81**:1809–1812.

84. **Nepom, J. T., M. Tardieu, R. L. Epstein, J. H. Noseworthy, H. L. Weiner, J. Gentsch, B. N. Fields, and M. I. Greene.** 1982. Virus-binding receptors: similarities to immune receptors as determined by anti-idiotypic antibodies. *Surv. Immunol. Res.* **1**:255–261.

85. **Nepom, J. T., H. L. Weiner, M. A. Dichter, M. Tardieu, D. R. Spriggs, C. F. Gramm, M. L. Powers, B. N. Fields, and M. I. Greene.** 1982. Identification of a hemagglutinin-specific idiotype associated with reovirus recognition shared by lymphoid and neural cells. *J. Exp. Med.* **155**:155–167.

86. **Noseworthy, J. H., B. N. Fields, M. A. Dichter, C. Sobotka, E. Pizer, L. L. Perry, J. T. Nepom, and M. I. Greene.** 1983. Cell receptors for the mammalian reovirus. I. Syngeneic monoclonal anti-idiotypic antibody identifies a cell surface receptor for reovirus. *J. Immunol.* **131**:2533–2538.

87. **Ohno, T., M. Nakamura, M. Kamada, M. Watanabe, Y. Kohno, and M. Kobayashi.** 1988. Complement dependent cytolysis of HIV infected cells with anti-idiotype monoclonal antibodies against the CD4 molecule. Abstract no. 2211. *IV International Conference on AIDS,* Stockholm. International Development Research Center, Ottawa, Canada.

88. **Paque, R. E., and R. Miller.** 1989. Polyclonal anti-idiotypes influence macrophage chemotaxis in coxsackievirus-induced murine myocarditis. *J. Leukocyte Biol.* **45**:79–86.

89. **Peterson, A., and B. Seed.** 1988. Genetic analysis of monoclonal antibody and HIV binding sites on the human lymphocyte antigen CD4. *Cell* **54**:65–72.

90. **Powell, T. J., Jr., R. Spann, M. Vakil, J. F. Kearney, and E. W. Lamon.** 1988. Activation of a functional idiotype network response by monoclonal antibody specific for a virus (M-MuLV)-induced tumor antigen. *J. Immunol.* **140**:3266–3272.

91. **Rath, S., V. Bal, B. J. Mohite, V. Haridas, S. R. Naik, S. A. Kamat, and A. J. Zuckerman.** 1988. Anti-idiotypic humoral and cellular responses to antibody to hepatitis B surface antigen in hepatitis B viral infections. *Clin. Exp. Immunol.* **73**:360–365.

92. **Reagan, K. J.** 1985. Modulation of immunity to rabies virus induced by anti-idiotypic antibodies. *Curr. Top. Microbiol. Immunol.* **119**:15–30.

93. **Reagan, K. J., W. H. Wunner, T. J. Wiktor, and H. Koprowski.** 1983. Anti-idiotypic antibodies induce neutralizing antibodies to rabies virus glycoprotein. *J. Virol.* **48**:660–666.

94. **Reale, M. A., A. J. Manheimer, T. M. Moran, G. Norton, C. A. Bona, and J. L. Schulman.** 1986. Characterization of monoclonal antibodies specific for sequential influenza A/PR/8/34 virus variants. *J. Immunol.* **137**:1352–1358.

95. **Rieber, E. P., H. Chen, C. Federle, and G. Fiethmuller.** 1988. Monoclonal anti-CD4-idiotype antibody recognizing the HIV-envelope gp120. Abstract no. 3089. *IV International Conference on AIDS,* Stockholm. International Development Research Center, Ottawa, Canada.

96. **Rossini, A. A., J. P. Mordes, and A. A. Like.** 1985. Immunology of insulin-dependent diabetes mellitus. *Annu. Rev. Immunol.* **3**:289–320.

97. **Saag, M. S., B. H. Hahn, J. Gibbons, Y. Li, E. S. Parks, W. P. Parks, and G. M. Shaw.** 1988. Extensive variation of human immunodeficiency virus type 1 in vivo. *Nature* (London) **334**:440–444.

98. **Sacks, D. L., and A. Sher.** 1983. Evidence that anti-idiotype induced immunity to experimental African trypanosomiasis is genetically restricted and requires recognition of combining site-related idiotopes. *J. Immunol.* **131**:1511–1515.

99. **Sattentau, Q. J., J. N. Weber, R. A. Weiss, and P. C. L. Beverley.** 1987. Antisera to Leu 3a with anti-idiotypic activity react with gp110/130 of HIV-1 and LAV-2. Abstract no. Th.9.4. *III International Conference on AIDS,* Washington, D.C. International Development Research Center, Ottawa, Canada.

100. **Schick, M. R., G. R. Dreesman, and R. C. Kennedy.** 1987. Induction of an anti-hepatitis B surface antigen response in mice by noninternal image (Ab2a) anti-idiotypic antibodies. *J. Immunol.* **138**:3419–3425.

101. **Sege, K., and P. A. Peterson.** 1978. Anti-idiotypic antibodies against anti-vitamin A transport protein reacts with prealbumin. *Nature* (London) **271**:167–168.

102. **Sege, K., and P. A. Peterson.** 1978. Use of anti-idiotypic antibodies as cell-surface receptor probes. *Proc. Natl. Acad. Sci. USA* **75**:2443–2447.

103. **Sharpe, A. H., and B. N. Fields.** 1981. Reovirus inhibition of cellular DNA synthesis: role of the S1 gene. *J. Virol.* **38**:389–392.

104. **Sharpe, A. H., G. N. Gaulton, H. C. J. Ertl, R. W. Finberg, K. K. McDade, B. N. Fields, and M. I. Greene.** 1985. Cell receptors for the mammalian reovirus. IV. Reovirus-specific cytolytic T cell lines that have idiotypic receptors recognize anti-idiotypic B cell hybridomas. *J. Immunol.* **134**:2702–2706.

105. **Sharpe, A. H., G. N. Gaulton, K. K. McDade, B. N. Fields, and M. I. Greene.** 1984. Syngeneic monoclonal antiidiotype can induce cellular immunity to reovirus. *J. Exp. Med.* **160**:1195–1205.

106. **Sigal, N. H., M. Chan, M. A. Reale, T. Moran, Y. Beilin, J. L. Schulman, and C. Bona.** The human and murine influenza-specific B cell repertoires share a common idiotope. *J. Immunol.* **139**:1985–1990.

107. **Silverton, E. W., M. A. Navia, and D. R. Davies.** 1977. Three-dimensional structure of an intact human immunoglobulin. *Proc. Natl. Acad. Sci. USA* **74**:5140–5144.

108. **Simons, K., and H. Garoff.** 1980. The budding mechanisms of enveloped animal viruses. *J. Gen. Virol.* **50**:1–21.

109. **Smorodinsky, N. I., Y. Ghendler, R. Bakimer, S. Chaitchuk, I. Keydar, and Y. Shoenfeld.** 1988. Towards an idiotypic vaccine against mammary tumors. Induction of an immune response to breast cancer-associated antigens by anti-idiotypic antibodies. *Eur. J. Immunol.* **18**:1713–1718.

110. **Staudt, L. M., and W. Gerhard.** 1983. Generation of antibody diversity in the immune response of BALB/c mice to influenza virus hemagglutinin. *J. Exp. Med.* **157**:687–704.

111. **Stein, K., and T. Soderstrom.** 1984. Neonatal administration of idiotype primes for protection against Escherichia coli K13 infection in mice. *J. Exp. Med.* **160**:1001–1011.

112. **Swain, S. L., D. P. Dialynas, F. W. Fitch, and M. English.** 1984. Monoclonal antibody to L3T4 blocks the function of T cells specific for class 2 major histocompatibility complex antigens. *J. Immunol.* **132**:1118–1123.

113. **Szmuness, W., C. E. Stevens, E. J. Harley, E. A. Zang, W. R. Olesko, D. C. William, R. Sadovsky, J. M. Morrison, and A. Kellner.** 1980. Hepatitis B vaccine. Demonstration of efficacy in a controlled clinical trial in a high-risk population in the United States. *N. Engl. J. Med.* **303**:833–841.

114. **Tanaka, M., N. Sasaki, and A. Seto.** 1986. Induction of antibodies against Newcastle disease virus with syngeneic anti-idiotype antibodies in mice. *Microbiol. Immunol.* **30**:323–331.

115. **Taub, R., R. J. Gould, V. M. Garsky, T. M. Ciccarone, J. Hoxie, P. A. Friedman, and S. J. Shattil.** 1989. A monoclonal antibody against the platelet fibrinogen receptor contains a sequence that mimics the receptor recognition domain in fibrinogen. *J. Biol. Chem.* **264**:259–265.

116. **Thanavala, Y. M., A. Bond, R. Tedder, F. C. Hay, and I. M. Roitt.** 1985. Monoclonal internal image anti-idiotypic antibodies of hepatitis B surface antigen. *Immunology* **55**:197–204.

117. **Thanavala, Y. M., S. E. Brown, C. R. Howard, I. M. Roitt, and M. W. Steward.** 1986. A surrogate hepatitis B virus antigenic epitope represented by a synthetic peptide and an internal image antiidiotype antibody. *J. Exp. Med.* **164**:227–236.

118. **Thanavala, Y. M., M. J. Moore, R. Tedder, F. C. Hay, and I. M. Roitt.** 1988. Recognition of several idiotopes on a monoclonal anti-hepatitis B virus surface antigen using monoclonal anti-idiotypic antibodies. *Immunology* **63**:575–577.

119. **Thorbecke, G. J., and G. W. Siskind.** 1984. Auto-anti-idiotype production during the response to antigen, p. 417–431. *In* M. I. Greene and A. Nisonoff (ed.), *The Biology of Idiotypes.* Plenum Publishing Corp., New York.

120. **Troisi, C. L., and F. B. Hollinger.** 1985. Detection of an IgM antiidiotype directed against anti-HBs in hepatitis B patients. *Hepatology* **5**:758–762.

121. **Urbain, J., P.-A. Cazenave, M. Wikler, J. D. Franssen, B. Mariame, and O. Leo.** 1980. Idiotypic induction and immune networks, p. 81–87. *In* M. Fougereau and J. Dausset (ed.), *Immunology 80: Progress in Immunology IV.* Academic Press, Inc. (London), Ltd., London.

122. **Urbain, J., M. Slaoui, B. Mariame, and O. Leo.** 1984. Idiotypy and internal images, p. 15–28. *In* H. Köhler, J. Urbain, and P.-A. Cazenave (ed.), *Idiotypy in Biology and Medicine.* Academic Press, Inc., New York.

123. **Uytdehaag, F. G. C. M., H. Bunschoten, K. Weijer, and A. D. M. E. Osterhaus.** 1986. From Jenner to Jerne: towards idiotype vaccines. *Immunol. Rev.* **90**:93–113.

124. **Uytdehaag, F. G. C. M., and A. D. M. E. Osterhaus.** 1985. Induction of neutralizing antibody in mice against poliovirus type II with monoclonal anti-idiotypic antibody. *J. Immunol.* **134**:1225–1229.

125. **Vaux, D. J. T., A. Helenius, and I. Mellman.** 1988. Spike-nucleocapsid interaction in Semliki Forest virus reconstructed using network antibodies. *Nature* (London) **336**:36–42.

126. **Weiner, D. B., W. V. Williams, J. A. Hoxie, J. A. Berzofsky, and M. I. Greene.** 1988. Non CD4 molecules on human cells important in HIV-1 human cell interactions. *Vaccines* **89**:114–119.

127. **Weiner, D. B., K. Hubner, W. V. Williams, and M. I. Greene.** 1989. Species trophism of HIV-1: infectivity of interspecific cell hybridomas implies non-CD4 structures are required for cell entry. *Cancer Detect. Prev.,* in press.

128. **Weiner, D. B., W. V. Williams, J. A. Hoxie, and M. I. Greene.** 1988. Identification of ancillary molecules on human T cells important to HIV-1 T cell interactions. Abstract no. 2577. *IV International Conference on AIDS,* Stockholm. International Development Research Center, Ottawa, Canada.

129. **Wilde, D. B., P. Marrack, J. Kappler, D. P. Dialynas, and F. W. Fitch.** 1983. Evidence implicating L3T4 in class II MHC antigen reactivity: monoclonal antibody GK1.5 (anti-L3T4a) blocks class II MHC antigen-specific proliferation, release of lymphokines, and binding by cloned murine helper T lymphocyte lines. *J. Immunol.* **131:**2178–2183.

130. **Williams, W. V., H. R. Guy, J. A. Cohen, D. B. Weiner, and M. I. Greene.** 1988. Molecular and immunologic analysis of a functional internal image formed by an anti-receptor antibody. *Ann. Inst. Pasteur Immunol.* **139:**659–675.

131. **Williams, W. V., H. R. Guy, D. H. Rubin, F. Robey, J. N. Myers, T. Kieber-Emmons, D. Weiner, and M. I. Greene.** 1988. Sequences of the cell-attachment sites of reovirus type 3 and its anti-idiotypic/antireceptor antibody: modeling of their three-dimensional structures. *Proc. Natl. Acad. Sci. USA* **85:**6488–6492.

132. **Williams, W. V., S. D. London, D. B. Weiner, S. Wadsworth, J. A. Berzofsky, F. Robey, D. H. Rubin, and M. I. Greene.** 1989. Immune response to a molecularly defined internal image idiotope. *J. Immunol.* **142:**4392–4400.

133. **Woodland, R., and H. Cantor.** 1978. Idiotype-specific T helper cells are required to induce idiotype-positive B memory cells to secrete antibody. *Eur. J. Immunol.* **8:**600–606.

134. **Zhou, E.-M., T. C. Chanh, G. R. Dreesman, P. Kanda, and R. C. Kennedy.** 1987. Immune response to human immunodeficiency virus: in vivo administration of anti-idiotype induces an anti-gp160 response specific for synthetic peptide. *J. Immunol.* **139:**2950–2956.

Chapter 4

Occurrence, Roles, and Uses of Idiotypes and Anti-Idiotypes in Parasitic Diseases

Daniel G. Colley

THE NATURE OF PARASITIC DISEASES

All infectious organisms can be thought of as parasitic in that they benefit from their host and usually offer little, or nothing positive, in return. However, the conventional use of the term parasitic disease is in reference to infections caused by eucaryotes of the animal kingdom, including protozoa, helminths, and ectoparasitic arthropods. This chapter will deal with idiotypic interactions that occur spontaneously during infections with these organisms or that have been induced in relationship to these organisms.

Why should the immunobiology of infections by these organisms be considered in a separate chapter? The historical segregation of studies of these infections, based on taxonomic differences and relative geographic separations, is not a compelling reason. Although the perception of parasites as dauntingly complex targets of immunologic interest, and their existence in sometimes exotic settings, set them apart, the concepts and approaches that must be applied to studies of these organisms do not differ from those of other immune systems.

There is, however, a differential aspect of most of these infections that is reason enough to focus on studies of idiotypic interactions in parasitic infections. Although most of these infections exhibit an acute phase, as a group they are usually seen as long-term, chronic infections. This often lifelong chronicity is fundamental to their immunobiology and is thought to

Daniel G. Colley • Veterans Administration Medical Center and Departments of Microbiology and Immunology, and Medicine, Vanderbilt University School of Medicine, Nashville, Tennessee 37212.

contribute to the likelihood of relevant idiotypic interactions (32). Chronicity requires that the immune system and the parasite have evolved a stable, balanced host-parasite relationship. In many of these infections this involves, and may be due to, a high degree of immunoregulatory interactions (27, 136). This immunoregulation can rest, in part, on idiotypic relationships (31, 32, 39). Furthermore, chronicity of an endemic infection through the child-bearing years means that the vast majority of individuals who are going to be infected by a given organism (those living in the endemic area) were gestated and often breast-fed by mothers who harbored that active, chronic infection. This means that almost all of those who will become infected during their lifetime were exposed in utero and neonatally to the idiotypes on their mothers' high-titer antiparasite antibodies (69, 89, 94, 95, 100) and probably the idiotypes on their mothers' anti-idiotypes to those antiparasite antibodies. They were also most probably exposed to parasite antigens circulating in their mothers' blood (47, 104) and/or present in milk (133). This situation offers multiple opportunities for in utero sensitization (152) and involvement of perinatal idiotypic and anti-idiotypic influences (46).

Another area of interest in idiotypic interactions in parasitic infections (128) is similar to that for other organisms, i.e., the possible use of anti-idiotypic vaccines (67). This approach to vaccines may be particularly needed in the armament against many parasites because some of the life cycle stages of these organisms are difficult to obtain in sufficient quantities to isolate potential immunogens. Also, as with some other infectious agents, the pertinent immunogens may be carbohydrates, preventing their production as recombinants.

Studies of idiotypes in parasitic diseases have included such infections as African trypanosomiasis, Chagas' disease, malaria, coccidiosis, lymphatic filariasis, onchocerciasis, and schistosomiasis. This chapter will review the findings of these studies and suggest some themes which may be common to some of the infections caused by these widely disparate organisms.

THE NATURE OF STUDIES OF IDIOTYPIC–ANTI-IDIOTYPIC INTERACTIONS

Studies of idiotypes and anti-idiotypic responses seem to focus on three main areas of endeavor: (i) studies of the gene and protein structural basis for idiotypy; (ii) theoretical and experimental studies of idiotypy as the foundation of an interactive immune system; and (iii) studies of the functional roles of natural or induced idiotypic interactions in situations such as ontogeny, allergy, neoplasia, autoimmunity, and infectious diseases. All three of these fields of study have yielded stimulating information concerned with the basis of the immune system. The third area is currently that which encompasses most studies of idiotypy in parasitic disease. However, some of the information being generated by idiotypic studies of parasitic infections may lead to insights into what occurs in the more fundamental second area, especially in regard to maternal and perinatal influences (46). None of the parasitic disease systems has yet been studied in enough idiotypic detail to contribute to the first area, concerning structural correlates among gene rearrangement, protein structure, and idiotopes. The identification of dominant, cross-reactive idiotypes (CRI) on antiparasite antibodies (85, 107, 108) has begun to demonstrate particular idiotypes of functional consequence in these infections. Continued study of such idiotypes, especially those that stimulate T cells directly in contrast to those that act through accessory-cell processing and presentation (30), may also contribute information on idiotypic structural relationships.

INDUCTION OF PROTECTIVE IMMUNE RESPONSES AGAINST PARASITES BY USE OF ANTI-IDIOTYPIC ANTIBODIES

The development of antiparasitic vaccines is a challenging enterprise (96, 135). These are mostly organisms which are well adapted to their hosts, and although they usually induce strong and varied immune responses, they are generally well tolerated by the host throughout long chronic infections. The selection of protective immunogens and appropriate regimens has proven difficult (135–137). The use of recombinant DNA technology, to overcome the production difficulties associated with organisms that are hard to grow in bulk, is currently the major focus of efforts to produce candidate parasite protein vaccines (28, 97, 117). However, many parasite immunogens, and some potential vaccine target antigens, may be carbohydrate portions of glycoproteins or glycolipids, or pure carbohydrates. It is possible that the use of mimicry by monoclonal antibodies (MAbs) which are anti-idiotypic against protective MAbs that react with the target carbohydrate epitope could induce adequate resistance.

African Trypanosomiasis

In 1982, Sacks et al. (129) were the first to demonstrate the potential of such antigen-independent induction of antimicrobial immunity by using anti-idiotypic antibodies. They were studying the parasitic infection African trypanosomiasis in mice, caused by *Trypanosoma rhodesiense*. They prepared the immunoglobulin G1 (IgG1) fractions of anti-idiotypic rabbit sera against three different protective MAbs which reacted with the protein portion of the variant-specific glycoprotein of a clone of *T. rhodesiense*. The

IgG1 fractions of these anti-idiotypic sera were given to mice 3 to 4 weeks before challenge with *T. rhodesiense*. This treatment either completely protected, reduced resultant parasitemias, or selected against parasites bearing the original variant-specific glycoprotein. The specificity of the resistance induced by the anti-idiotypic immunoglobulins was clear, based on the ability of a given anti-idiotype to alter the course of infection of only the homologous clone of *T. rhodesiense*. The ability of these trypanosomes to alter their variant-specific glycoproteins hundreds of times by the process of antigenic variation (37) makes the usefulness of this particular anti-idiotypic vaccine a moot point, but the demonstration of the ability to induce an immune response in the absence of antigens of the infecting organism was of considerable scientific import. The authors went on to demonstrate (131) that major differences existed in the ability of their various anti-idiotype reagents to induce immunity. One of their Ab_2 preparations was an ineffective vaccine because it failed to induce the expression of the corresponding idiotype before challenge infection. Because Ab_3s developed only after infection, the organism established itself before protective levels were attained. Another of the Ab_2 preparations did induce high levels of Ab_3, but the response conferred no protection. This Ab_2 apparently induced a non-antigen-binding but idiotype-bearing set of Ab_3s. Immunity was induced by one of the Ab_2 preparations, and this resistance was restricted to mice bearing genes linked to Igh-C^a. Expression of the idiotype in question was apparently controlled by Igh-C^a in both this immunization regimen and the actual infection.

Schistosomiasis

In 1982, Grzych et al. (58) reported the production of a rat IgG2a MAb against *Schistosoma mansoni* schistosomula (the life cycle stage of schistosomes which is found in the mammalian host immediately after infection by cercariae) that participated in antibody-dependent cell-mediated cytotoxicity of schistosomula with eosinophils in vitro and led to high levels of resistance upon passive transfer to rats prior to challenge with *S. mansoni* cercariae. This MAb was specific for a carbohydrate epitope of a glycoprotein, schistosomular membrane antigen, of relative molecular mass 38 kilodaltons (kDa) (35). This epitope was immunogenic in several host species, including humans (36). The same epitope stimulated the production of rat IgG2c (59) and human IgM (81), blocking antibodies which interfered with in vitro and in vivo resistance-associated events.

Immunization of rats with the rat IgG2a-protective MAb induced an anti-idiotypic MAb (Ab_2) that inhibited the binding of Ab_1 to its 38-kDa target antigen (60). Sera from rats immunized with the purified anti-idiotypic (Ab_2) MAb contained specific anti-38-kDa schistosomular antigen

Ab$_3$ antibodies. These Ab$_3$s participated in eosinophil-mediated antibody-dependent cell-mediated cytotoxicity reactions against schistosomula and protected against *S. mansoni* challenge infections on passive transfer. The rats that were actively immunized with only Ab$_2$ MAb and whose sera contained the anti-schistosomular Ab$_3$ antibodies were likewise partially resistant (50 to 80%) (60). These studies demonstrated the use of an Ab$_2$ specific for an Ab$_1$ against a parasite carbohydrate epitope for effective protective immunization.

Other anti-idiotypic studies of the rat-*S. mansoni* resistance system have recently focused on a protective rat IgE MAb against an epitope present on schistosomular antigens of 26 and 56 kDa (149). This IgE Ab$_1$ induced anti-idiotypic Ab$_2$ antibodies and led to the production of anti-anti-idiotypic Ab$_3$ antibodies. The Ab$_3$ antibodies were of IgG and IgE isotypes, and both mainly recognized the 26-kDa schistosomular antigen. The IgG Ab$_3$ antibodies participated in vitro in eosinophil-mediated antibody-dependent cell-mediated cytotoxicity against schistosomula, and the IgE Ab$_3$ antibodies participated in vitro in platelet-mediated antibody-dependent cell-mediated cytotoxicity against schistosomula. Partial resistance was passively transferred with both Ab$_3$ populations. The investigators speculated about the possibility that immunization with this IgE Ab$_1$ antibody influenced the isotype (again IgE) of some of the Ab$_3$ antibodies produced upon immunization.

In the mouse system, heterologous (rabbit) anti-idiotypic antibodies against a protective anti-*S. mansoni* IgM MAb also induced partial resistance against cercarial challenge (86). The protective IgM MAb was specific for a 68-kDa glycoprotein in adult worm extracts and schistosomular membranes (83). Immunization of one of three rabbits with this MAb produced anti-idiotypic (Ab$_2$) antibodies. Immunization of mice with these rabbit antibodies led to partial resistance, even without the use of adjuvant. The immunized mice developed Ab$_3$ antibodies that bound to surface membranes of schistosomula. A portion (3 to 32%) of the induced humoral immune response bound to the original 68-kDa antigen.

Chagas' Disease

The potential for the use of the molecular mimicry approach of simulating a protozoan carbohydrate epitope has been demonstrated in experimental Chagas' disease (American trypanosomiasis), which is caused by *Trypanosoma cruzi* (130). Anti-idiotypic (Ab$_2$) antibodies were produced in rabbits immunized with an IgG2a mouse MAb against the carbohydrate portion of a 72-kDa surface glycoprotein (GP72) of epimastigotes and metacyclic trypomastigotes of *T. cruzi* (142). Immunization of mice with this GP72 leads to their partial protection against infection with metacyclic trypomastigotes (141). The rabbit Ab$_2$ antibodies reacted with parasite-induced anti-*T. cruzi* antibodies from rabbits, mice, and humans.

Sacks et al. (130) used the rabbit Ab$_2$ antibodies they raised against the GP72-specific MAb to immunize mice, rabbits, and guinea pigs. The immunized animals responded by producing specific Ab$_3$ antibodies. Immunization of mice with the Ab$_2$ led to greatly enhanced responses of the original idiotype of the MAb upon subsequent *T. cruzi* infection. Thus, exposure to Ab$_2$ effectively primed mice for enhanced expression of idiotype-bearing, antigen-specific clones on antigenic exposure. However, this did not lead to active protection against *T. cruzi* challenge infections. Because passive transfer of the idiotype alone is also not protective, these negative data probably indicate only that in this particular system, responses against the carbohydrate portion of GP72 are insufficient to afford resistance. Nevertheless, the studies emphasize the ability to mimic a carbohydrate epitope by use of Ab$_2$ antibodies against a MAb specific for the carbohydrate epitope.

Coccidiosis

Coccidiosis, an infection of the intestinal epithelial cells that is caused by the protozoan *Eimeria tenella,* is a widespread commercial threat to the worldwide poultry industry. Because fowl that recover from this disease are resistant to reinfection, there is considerable interest in the development of an anticoccidial vaccine. Bhogal et al. (7) raised rabbit polyclonal Ab$_2$ anti-idiotypic antibodies with antiparatopic specificities against two protective mouse MAbs to *E. tenella* sporozoites. Immunization of young chickens with these Ab$_2$ antibodies induced antisporozoite antibodies and partial protective immunity to subsequent challenges with *E. tenella* sporozoites. Immunizations with anti-idiotypic antibodies against two other, nonprotective anti-*E. tenella* MAbs failed to protect chickens.

EXPRESSION OF AUTO-ANTI-IDIOTYPIC RESPONSES IN PARASITIC DISEASES

Many parasitic infections result in an acute (sometimes symptomatic) stage of disease, which then develops into a chronic (again, sometimes symptomatic) stage. The acute phase of many of these infections (especially African and American trypanosomiasis and malaria) is associated with polyclonal B-cell activation and immune complex-like symptomatology. Chronic parasitic infections commonly last many years, often a lifetime, and are associated with continued antibody production and T-cell responsiveness against antigens of the infecting organism. Strong responses often continue, even when it is difficult to demonstrate organisms in the host.

During the polyclonal B-cell activation observed in BALB/c mice acutely infected with the African trypanosome *T. brucei* subsp. *brucei,* it has been seen that both anti-phosphorylcholine antibody-producing plaque-forming cells (PFC) making antibodies bearing the T15 idiotype and PFC making anti-T15 anti-idiotype antibodies occurred simultaneously, 6 to 12 days after initial infection (125). From day 11 after infection, immune complexes were found in the sera of these mice. It was conjectured that the polyclonal stimulation of B cells in such acute parasitic disease settings could be responsible for idiotype–anti-idiotype immune complexes and contribute to the observed disease manifestations. Parallel findings were reported to occur following *Plasmodium yoelii* infections in BALB/c mice (88).

The chronic, 24-hour-a-day stimulation of antiparasitic responses, due to repeated (sometimes continuous) production and release of parasite antigens, would seem to provide excellent opportunities for anti-idiotypic responses against the chronically expressed idiotypes. The long-term presence of the antibodies and T-cell receptors specific for parasite epitopes may be central to the development of conspicuous auto-anti-idiotypic responses against them.

Schistosomiasis

Experimental model studies

During most of their adult lives, schistosomes live in the blood vessels around the gut (*S. mansoni* and *S. japonicum*) or bladder (*S. haemato-bium*). There the adult male and female worms mate, more or less continuously, and produce fertilized eggs. Some of the eggs escape from the body via feces or urine and perpetuate the life cycle. However, many eggs are swept by the blood to the presinusoidal capillaries of the liver, where they lodge and release strongly immunogenic soluble egg antigens (SEA). These antigens induce humoral and cellular anti-SEA responses soon after egg production begins, 4.5 to 5 weeks after infection (23, 118).

With a modified reverse PFC assay, Powell and Colley (123) demonstrated that splenic auto-anti-idiotypic PFC, making antibodies specific for anti-SEA antibodies, develop in mice by 8 weeks after *S. mansoni* infection. Thus, by 3 weeks after the onset of egg production and about 1 to 1.5 weeks after anti-SEA antibodies are detected in the serum, specific anti-idiotypic PFC are detectable in the spleens of infected mice. These anti-idiotypic PFCs are found thereafter and can be positively selected by cell panning on plates coated with anti-SEA antibodies. When individual mice, infected for 8 to 16 weeks, were evaluated for the levels of anti-SEA antibodies in their sera (by SEA-specific enzyme-linked immunosorbent assays [ELISAs]) and those levels were compared with their individual numbers of anti-anti-SEA (anti-idiotypic) PFC per spleen for each mouse, a strong inverse correlation was observed (32). Thus, a mouse that expresses high levels of circulating anti-SEA antibodies can be predicted ($P < 0.001$) to have low levels of anti-anti-SEA PFC in its spleen, and vice versa. This reciprocal idiotype/anti-idiotype relationship, in the continuous presence of the initiating antigen (SEA from continually produced eggs), could be explained by the induction of sequential waves of autologously generated idiotypic and anti-idiotypic antibodies or cells (14, 17, 18, 34, 52, 57, 79, 80, 134).

The development of auto-anti-idiotypic $Thy1^+$ $CD4^+$ $Lyt2^-$ lymph node cells specifically responsive to anti-SEA antibodies also occurs by 8 weeks after *S. mansoni* infections of mice (124). These studies were patterned after the demonstration of such auto-anti-idiotype-reactive T cells in the peripheral blood of patients with schistosomiasis mansoni (91) (see below).

Olds and Kresina (107) demonstrated the development of auto-anti-idiotypic antibodies against anti-SEA antibodies in the mouse-*S. japonicum* model. Their studies examined the specificity of the anti-idiotypic antibodies expressed at different times after infection. Anti-idiotypes isolated from the sera of mice infected for 12 weeks identified a minor CRI ($SJ\text{-}CRI_m$), whereas the anti-idiotypes found in sera from mice infected for 30 weeks reacted with a major CRI ($SJ\text{-}CRI_M$), found on >50% of the anti-SEA antibodies. The presence of auto-anti-idiotypic networks in chronic, experimental schistosomiasis seems solidly established for both *S. mansoni* and *S. japonicum* infections.

Auto-anti-idiotypic antibody responses have also been demonstrated in the mouse-*S. mansoni* system in regard to protective humoral responses against soluble cercarial immunogens. Using antigen/paratope-specific blocking ELISA systems, Phillips et al. (119) showed that mice infected with *S. mansoni* for 8 to 10 weeks develop circulating specific anti-idiotypic antibodies against two different protective anti-soluble cercarial immunogen MAbs that have different epitopic specificities. The ability to remove these inhibiting activities by absorption on affinity columns made of only the homologous MAb, and not other potentially related columns, defined the nature and specificity of these auto-anti-idiotypes.

Other idiotypic studies, concerned with a protective MAb that is specific for a component of SEA and a schistosomular membrane antigen, have shown that some of the hybridomas made from the spleens of mice immunized to the MAb (Ab_1) produced Ab_2 antibodies which inhibited the binding of Ab_1 to antigen (113). It is also of considerable interest that other hybridomas derived from the same Ab_1-immunized mice produced antibodies that bound a component of SEA and were inhibited in their ability to do so by some of the Ab_2 MAbs. These were thus considered Ab_3s, providing evidence of the feasibility of Ab_1-Ab_2-Ab_3 networks in schistosomiasis.

Human studies

As indicated above, auto-anti-idiotypic T lymphocytes can be demonstrated in peripheral blood mononuclear cells (PBMC) of patients with schistosomiasis mansoni (91). These cells proliferate when cultured in the presence of immunoaffinity-purified anti-SEA antibodies (91) from the same (autologous) patient, another patient, or a pool of patient sera. Normal immunoglobulin preparations are not stimulatory to these PBMC. This occurs with PBMC from schistosomiasis patients or former patients, but not people who have never had schistosomiasis. The responding cells are T cells. PBMC from some but not all patients respond to any given anti-SEA antibody preparation, and cells from an individual are less likely to respond to anti-SEA antibodies from another patient than those from a serum pool (91). Goes et al. (56) have made human heterohybridomas which produce human anti-SEA MAbs that can stimulate the proliferation of PBMC from some patients or former patients (32, 33). Also, some mouse anti-SEA MAbs can stimulate PBMC from some patients and former patients (31). At the same time, anti-SEA preparations from certain patient groups do not stimulate PBMC from chronically infected patients or former patients. For example, anti-SEA antibodies from infected children under the age of 13 years or patients with acute schistosomiasis (symptomatic and of approximately 3 months duration) were not stimulatory, for cells from either the antibody donors or other patients (32). These nonstimulatory anti-SEA antibodies were prepared over the same SEA affinity columns, and their anti-SEA titers in ELISAs were the same as those of stimulatory anti-SEA antibodies prepared in parallel from sera of other given groups of chronically infected patients.

The working hypothesis in the above studies has been that idiotypes on the stimulatory anti-SEA antibodies (polyclonal or monoclonal) are responsible for the induction of T-cell proliferation. There is no evidence of any SEA in the antibody preparations, and if contaminating SEA is present it seems not to be responsible for the stimulation observed. This conclusion is supported by the parallel production of nonstimulatory antibodies, by the activity of MAbs, and by data that indicate that the mode of stimulation of this T-cell response is direct, i.e., that it occurs without a need for antigen-processing or major histocompatibility complex (MHC) presentation (see below). The specificity of anti-SEA antibody-mediated stimulation of PBMC of schistosomiasis patients was seen in studies in which anti-epimastigote antibodies, isolated by immunoaffinity over columns made of *Trypanosoma cruzi* epimastigote antigens (EPI), did not stimulate PBMC from patients with schistosomiasis (51). The reciprocal was also true; i.e., anti-SEA antibodies did not stimulate the PBMC from patients with Chagas' disease (51). PBMC from patients with Chagas' disease were stimulated to proliferate when cultured with the anti-EPI immunoaffinity-purified antibodies (50, 51).

Studies of auto-anti-idiotypic T cells from patients with schistosomiasis showed that either PBMC or purified T cells plus adherent cells responded to the stimulatory anti-SEA antibodies, but purified T cells, adherent cells, and cells enriched for B cells did not (91). Further studies were pursued regarding the mode of stimulation of T cells by these antibodies and the role of adherent cells in this process. Chloroquine, either during a prepulse of the adherent cells with antibodies or throughout the cultures (at 20 to 25 μM), did not alter (or increased [see below]) the response of the T cells (111). In parallel cultures the responses of the same PBMC to SEA were virtually eliminated (mean decrease = 90%) by these chloroquine procedures. These data are interpreted to mean that stimulation by anti-SEA antibodies does not require processing of the antibodies by antigen-presenting cells, represented by the adherent cells in this system. As expected, stimulation by SEA does require a chloroquine-sensitive processing step. The Fc portion of stimulatory anti-SEA antibodies was not required for stimulation, since $F(ab')_2$ fragments were completely stimulatory to PBMC (but not purified T cells) of the patients. Their purified T cells could, however, respond to $F(ab')_2$ fragments if the cultures also received exogenous interleukin-1 (IL-1) (purified or recombinant). This stimulation by soluble $F(ab')_2$ plus exogenous IL-1 was inhibited by the addition of anti-IL-1 rabbit serum but not normal rabbit serum. Soluble Fab fragments of the stimulatory anti-SEA antibodies were not stimulatory. However, when the same, nonstimulatory Fab fragments were coupled to Sepharose 4B beads, they stimulated purified T cells in the presence of exogenous IL-1. Thus, the Fab fragment of the stimulatory anti-SEA antibodies contains the stimulatory sequence or shape, but this signal must be delivered either bivalently or on a matrix, and in the presence of IL-1, if it is to lead to anti-idiotypic T-cell proliferation. The data indicate that this particular triggering of auto-anti-idiotypic T cells occurs directly through a cross-linking interaction of the stimulatory idiotype and the stimulated T cell. At this time there is no formal proof that this interaction occurs through T-cell receptors, but the specificity of the interaction and the observations described above make this the most likely choice. This direct mode of idiotype recognition by, and stimulation of, anti-idiotypic T cells, without processing and in the absence of MHC antigens, is not unique to this system (11, 12, 18, 19, 42–44, 74, 86). However, neither is it the only way such stimulation of anti-idiotypic T cells occurs (71), because some T cells respond to idiotypes in an MHC-restricted manner (8, 55, 62, 72, 73, 76, 77, 148). In the human schistosomiasis system, all of the multiclonal anti-SEA antibody preparations evaluated to date that stimulated PBMC from patients or former patients did so in a manner that was insensitive to the effects of chloroquine.

Chagas' Disease

As mentioned previously, PBMC from patients with Chagas' disease respond to anti-EPI antibodies, but not anti-SEA antibodies (51). Also, PBMC from some patients unfortunate enough to have both of these chronic parasitic diseases are stimulated by both anti-EPI and anti-SEA antibody preparations (51). The ease with which it is demonstrated that patients with either of these chronic, endemic (but otherwise very dissimilar) infections develop specific auto-anti-idiotypic T cells could suggest that the similarities of these infections (chronicity and endemicity) play a critical role in the establishment of this situation.

As in schistosomiasis, idiotypic stimulation of auto-anti-idiotypic T cells from patients with Chagas' disease required adherent-cell participation. Studies of this requirement in the Chagas' disease system have demonstrated that the mode of idiotypic stimulation is more complex than that studied in schistosomiasis with anti-SEA antibodies. This complexity became apparent when studies were done with anti-EPI antibodies derived from sera which were pooled based on the donor's clinical form of Chagas' disease. The responses of PBMC from patients with the severe, cardiac, or digestive form of Chagas' disease exposed to anti-EPI antibodies from the same group of patients (anti-EPI from cardiac patients, referred to as IdC) were not inhibited by chloroquine (30, 49a). The PBMC from these cardiac chagasic patients (with severe disease) can also be stimulated by anti-EPI antibodies from pooled sera of asymptomatic chagasic patients (the so-called indeterminate form of Chagas' disease; their anti-EPI antibodies being termed IdI). However, in direct contrast to the chloroquine-insensitive stimulation of their PBMC by IdC anti-EPI antibodies, the stimulation of their cells by IdI anti-EPI antibodies was almost completely suppressed in the presence of chloroquine. As expected, stimulation of PBMC from all chagasic patients by the EPI antigens from *T. cruzi* was uniformly suppressed by chloroquine. In addition, the responses of PBMC from patients with indeterminate (asymptomatic) Chagas' disease to both IdC and IdI were also chloroquine sensitive. More detailed analysis of the direct (chloroquine-insensitive) IdC-stimulated responses of T cells from cardiac Chagas' disease patients showed that the adherent-cell requirement of these responses could be met by the addition of exogenous IL-1. Also, neither anti-human leukocyte antigen (HLA)-DR,DP(DQ) MAb nor 0.0025% sodium azide inhibited the IdC-stimulated T-cell proliferation of patients with severe disease (49a). However, anti-HLA-DR,DP(DQ) MAb or 0.0025% sodium azide completely inhibited their T-cell responses to IdI or EPI antigen. This complex picture can be summarized (Table 1) by saying that patients with severe disease have auto-anti-idiotypic T cells that respond directly to their own anti-EPI antibodies (i.e., IdC). They also have auto-anti-idiotypic T cells

Table 1. Severe Clinical Disease in Chagasic Patients Who Express Directly Stimulatory Idiotypes on Their Anti-EPI Antibodies and Whose Cells Respond to These Idiotypes Directly

	Patient clinical form			
	Indeterminate		Cardiac	
Proliferative response[a] to:	Processed and presented[b]	Direct[c]	Processed and presented	Direct
T. cruzi				
Antigens	+++	---	++	---
IdI	++	---	+++	---
IdC	+++	--- [d]	+??+[e]	++++ [d]

[a] Response levels indicated by pluses, from low (+) to high (++++) levels of ^{3}H-TdR incorporation.
[b] Response of PBMC in the absence of chloroquine or antibodies against class II MHC antigens.
[c] Response of PBMC in the presence of chloroquine or antibodies against class II MHC antigens or purified T lymphocytes with exogenous IL-1.
[d] The boxes denote responses that separate the symptomatic and asymptomatic patients by more than just degree of responsiveness. (From reference 30 and studies in progress.)
[e] The proportion of this response due to stimulation by processed and preserved idiotypes is not defined.

against the idiotypes on anti-EPI antibodies from asymptomatic patients, but to respond to these IdI idiotypes, the idiotypes must be processed and presented in the context of the MHC. Patients with asymptomatic Chagas' disease also develop auto-anti-idiotypic T cells, but for these to respond to either IdC or IdI idiotypes, the idiotypes must be processed and presented in the context of class II MHC antigens (49a). Speculation about the relationships of these observations on auto-anti-idiotypic T cells to clinical disease is discussed in Chagas' Disease: Correlations of Idiotypic–Anti-Idiotypic Systems with Clinical Forms in Humans (see below).

Another recent report of auto-anti-idiotypic responses in Chagas' disease also indicates a relationship with the clinical form of the infection. Using F(ab′)₂ fragments of immunoglobulin from pooled sera of infants with acute Chagas' disease and overt parasitemia as their source of idiotype, Sadigursky et al. (132) showed that sera of 20 of 23 patients with cardiac Chagas' disease reacted with these F(ab′)₂ fragments. Only 30 of 92 sera from patients with indeterminate Chagas' disease bound these preparations, and only 4 of 84 sera from nonchagasic controls did so. These studies demonstrate the presence and correlation of auto-anti-idiotypic antibodies in severe chronic Chagas' disease, which are specific for idiotypes on immunoglobulins present in acute disease. As with the anti-idiotypic T-cell studies, the role(s) of these auto-anti-idiotypic responses in the control or pathogenesis of Chagas' disease remains conjectural.

REGULATION

Highly specific regulation of both the development and expression of the immune repertoire is one of the most closely scrutinized aspects of the immunobiology of idiotypic and anti-idiotypic responses (9, 53, 84, 112). Many studies indicate that both positive regulation (augmentation) and negative regulation (suppression) of an immune response can be due to anti-idiotypic involvement (9, 78, 112). As discussed above, auto-anti-idiotypic responses have been found in several parasitic infections and could positively or negatively influence different host-parasite interactions. In schistosomiasis and Chagas' disease, correlations have been seen between various auto-anti-idiotypic responses and expressions of either high versus low resistance or asymptomatic versus pathogenic disease. The relationships have led to considerations of whether idiotypic and anti-idiotypic interactions result from, control, or merely correlate with these events.

Immunopathology

Schistosomiasis

In endemic areas, schistosomiasis is seen primarily as a chronic, generally well-tolerated infection (27, 75, 103). Most of the infected population in endemic areas never experience clinical acute schistosomiasis (54, 103, 150), and only a relatively small proportion (5 to 10%) of patients ever exhibit the severe, periportal fibrosis and hepatosplenism which can develop in schistosomiasis mansoni or schistosomiasis japonica or the extreme urogenital fibrotic lesions attributable to schistosomiasis haematobium. In experimental studies of *S. mansoni* infection in mice, morbidity is largely attributable to cellular immune reactivities against schistosome SEA (118, 151). These responses against the SEA released from schistosome eggs trapped in the liver produce granulomas and are thought to contribute to the resultant fibrosis. This SEA-induced immunopathogenesis is also known to be regulated by anti-SEA-specific immunoregulatory responses, which develop coordinately with the establishment of chronic infection (4) and are mediated primarily by cellular responses (10, 24–26, 118, 120).

Studies in experimental models. In the mouse-*S. mansoni* model, a connection between the anti-SEA autoimmunoregulation and idiotypic/anti-idiotypic reactions was forged when it was found that a thymus-derived T-suppressor factor (TsF) obtained from anti-SEA-regulated, chronically infected mice that was capable of modulating (down regulating) schistosome egg-induced granuloma formation bound specifically to columns of anti-SEA antibodies, but not normal mouse immunoglobulin or SEA (2). This was taken as evidence that this *S. mansoni* T-suppressor factor (SmTsF) was anti-idiotypic for idiotypes expressed on anti-SEA antibodies from the sera of mice with chronic infections. Its characteristics suggested that this activity belonged to the group of T-suppressor factors known as TsF2s (5, 38). Other TsF activities have also been found in this anti-SEA, granuloma regulation circuit, and some have SEA-binding characteristics rather than binding to anti-SEA (22, 93, 114–116).

The situation with *S. japonicum* infection in mice is somewhat different, in that immunoregulation of the granulomatous pathogenesis also occurs through humoral mechanisms (49, 99, 109, 139, 143). Evidence for the involvement of idiotypic specificities in these regulatory responses is very strong. Regulatory roles have been assigned to both anti-SEA and auto-anti-anti-SEA antibodies (107). However, disparate effects on granuloma formation following immunizations designed to induce anti-idiotypic regulation to different anti-schistosomal MAbs demonstrate that it is difficult to predict which idiotypes are critically involved (99). The anti-idiotypic antibodies specific for a major CRI described by Olds and Kresina (SJ-CRI$_M$) (107), found in sera of chronic, well-regulated, 30-week-infected mice (described in Schistosomiasis: Experimental Model Studies), profoundly suppressed both in vitro anti-SEA responses and in vivo anti-egg granuloma formation. Anti-idiotypic antibodies specific for the serologically distinct minor CRI on anti-SEA antibodies from the sera of 12-week-infected mice, in the midst of the establishment of dominating regulatory responses, were not regulatory (107). Subsequent studies (85) showed that SJ-CRI$_M$ expression by antibodies decreased as the titer of anti-SEA antibody increased during infection. Examination of lymphoid cell populations for expression of receptors that express the SJ-CRI$_M$ idiotype found that in acute-stage infection (5 to 10 weeks postinfection) in the spleen, only B lymphocytes expressed SJ-CRI$_M$. However, in the thymus, SJ-CRI$_M$-bearing T lymphocytes were present from early in acute infections, until 15 weeks postinfection (85). More recently, Olds and Kresina (108) have generated multiple hybridomas from infected or SEA-immunized mice and selected them for the production of MAbs that suppress SEA-induced blastogenesis and/or anti-SEA or anti-anti-SEA activity. They then characterized these MAbs for their expression of SJ-CRI$_M$ and their antigen-binding specificities. Most of the immunosuppressive anti-SEA MAbs expressed SJ-CRI$_M$ and were derived from naturally infected mice, not those hyperimmunized to SEA. It is of considerable interest that all SJ-CRI$_M$-expressing anti-SEA MAbs gave identical banding patterns on immunoblots of SEA, and their binding to these SEA components was abrogated by sodium periodate treatment of the SEA. These studies strongly indicate that expression by antibodies of SJ-CRI$_M$ is associated with immunoregulation and that this immunoregulation is induced by a carbohydrate epitope(s) on SEA components.

The expression of SJ-CRI$_M$ on antibodies and cells was strongest early in infection and decreased as regulation of anti-SEA responses (including granuloma formation) and anti-SJ-CRI$_M$ idiotypic responses developed progressively (85). The correlation of SJ-CRI$_M$ expression, or the response to it, with immunoregulation, and the distinct and unexpected localization in the thymus of T cells bearing SJ-CRI$_M$, are of interest. In the *S. mansoni* system, the best cell-for-cell source of the anti-idiotypic, immunoregulatory extract termed SmTsF was the thymus (45). Idiotypic–anti-idiotypic interactions within the thymus are not well understood, but have the potential of exerting strong influences on the development and expression of subsequent immune responses (72).

Correlations of idiotypic–anti-idiotypic systems with clinical forms in humans. Auto-anti-idiotypic T cells can be demonstrated in schistosomiasis mansoni patients by their proliferation upon culture in the presence of anti-SEA antibodies (91). Montesano et al. (101) showed that the idiotypes expressed by immunoaffinity-purified anti-SEA antibodies differed, based both on their ability to stimulate in this proliferative system and on their recognition by specific rabbit anti-idiotypic sera. These idiotypic differences correlated very closely with the clinical form of schistosomiasis mansoni of the patients from whose sera they were derived.

Patients with schistosomiasis mansoni (and schistosomiasis japonica) are, on the basis of their history and clinical presentation, often diagnosed as having acute, intestinal (INT), hepatointestinal (HI), or hepatosplenic (HS) disease. The chronic, largely asymptomatic INT form is referred to as intestinal because although the worms live in mesenteric blood vessels, the only symptom experienced by the patients is periodic intestinal distress, thought to be associated with long-term excretion of eggs through the gut wall. In earlier studies it was observed that SEA-stimulated proliferative responses of PBMC from patients with the mild INT form were lower than those of ambulatory HS patients (29). A wide variety of regulatory responses have been demonstrated with the PBMC of INT patients (reviewed in reference 27), and some of these are lacking in HS patients (147). From studies by several different research teams, it appears that, in general, a chronic lack of regulation of anti-SEA T-cell responses is associated with the development of the severe, debilitating HS form of the disease (27, 103, 136). In contrast, well-regulated, lower, SEA-stimulated T-cell responses correlate with the largely asymptomatic INT presentation. The finding that anti-SEA antibodies purified from sera pooled from people with the INT

Table 2. Severe Hepatosplenic Disease in Schistosomiasis Patients Who Do Not Respond to Stimulatory Idiotypes Expressed on Anti-SEA Antibodies by an Immunoregulatory Response and Who Have Unregulated Responses to SEA[a]

Proliferative or immunoregulatory response[b] to:	Patient clinical form	
	Intestinal	Hepatosplenic
Proliferative		
SEA	++	++++ [c]
IdINT	++++	+++
IdHS	---	---
IdAcute	---	---
Immunoregulatory		
SEA	++++	++
IdINT	++++	---

[a] From references 39 and 101.
[b] Response level or degree of regulation indicated from low (+) to high (+++) levels.
[c] The boxes denote comparisons that suggest that regulatory responses, and particularly those induced by certain anti-SEA antibodies, are associated with being asymptomatic.

form differ idiotypically from those isolated from sera of patients with the acute or HS form (101) has added to the immunologic differences reported in these clinical presentations (Table 2).

Anti-SEA antibodies from INT or HI patients were strongly stimulatory for PBMC cells from patients with INT, HI, or HS disease. Anti-SEA antibodies from patients with acute or HS disease (although comparable in terms of anti-SEA activity, by both ELISA and immunoblot) did not stimulate the proliferation of PBMC from patients with any of these forms (101) (Table 2). Also, in competitive ELISAs, specific rabbit anti-idiotypic sera, raised against anti-SEA antibodies from acute, INT, HI, or HS patients, distinguished between the homologous anti-SEA antibodies (or other pools of anti-SEA antibodies from the same clinical group of patients) and anti-SEA antibodies from other clinical groups (101). On the basis of percent maximum inhibitions, this differential recognition was virtually 100% for comparisons between INT and acute and between HI and acute disease. Anti-SEA antibodies from INT patients shared some idiotypes (25 to 35%) with anti-SEA antibodies from HS patients and more (55 to 65%) with those from HI patients. Between two different groups of adult INT patient antibodies, the commonality of idiotypes was 75 to 85%. A discrepancy between

INT groups was observed when serum pools were based on the age of the INT patients (32). Anti-SEA antibodies isolated from two different pools of sera from INT children (younger than 13 years) failed to stimulate PBMC from either children or adults with INT infections. However, the PBMC from such children responded when cultured with anti-SEA antibodies from adult INT patients. This observation remains unexplained, but may indicate that children with INT disease fail to express stimulatory idiotypes on their anti-SEA antibodies, even though their cells are responsive to stimulatory idiotypes.

Further consequences of anti-SEA idiotype-stimulated T-cell proliferation have been studied with an in vitro granuloma system (6, 40, 41), which involves interactions of PBMC or purified populations of cells with SEA-coated beads. When cells from an INT patient were incubated for 24 h with stimulatory anti-SEA antibodies, washed, and added to autologous cells in an in vitro granuloma assay, there was a suppression of the developing granulomatous reaction (39, 40). Such preexposure of PBMC from decompensated HS patients to stimulatory anti-SEA antibodies from INT patients led to proliferation (101), but these proliferating cells did not exert a regulatory effect on in vitro granuloma formation (32, 39) (Table 2). Further studies of the regulatory responses induced by exposure of cells from patients with INT disease to stimulatory idiotypes (in this case expressed by several human anti-SEA MAbs) demonstrated that both CD4$^+$ and CD8$^+$ cells responded to the idiotypes. Cells of either phenotype, when stimulated by the idiotypes, interacted with autologous unfractionated PBMC to suppress anti-SEA in vitro granuloma formation (39). However, when only idiotype-prestimulated CD4$^+$ cells were mixed with purified autologous CD4$^+$ cells, no suppression resulted, but if the idiotype-precultured CD4$^+$ cells were mixed with autologous CD8$^+$ cells, granuloma formation was suppressed. Idiotype-prestimulated CD8$^+$ cells could directly suppress the granulomatous response of autologous cells (39). The data indicate that some idiotypes, expressed on multiclonal or monoclonal human anti-SEA antibodies, can stimulate in vitro suppressor-inducer/suppressor-effector circuitries that lead to the regulation of in vitro granuloma formation. Such immunoregulatory idiotypes are expressed primarily on anti-SEA antibodies from the asymptomatic INT patients with well-regulated anti-SEA responses, and not on anti-SEA antibodies from poorly regulated HS patients (Table 2).

Cross-reactive idiotypes shared between humans and mice. When rabbit anti-idiotypic sera raised to human anti-SEA antibodies (101) were used in competitive ELISAs, sera from mice infected with *S. mansoni* competed with the stimulatory CRIs from INT patients described above, but not with the nonstimulatory anti-SEA antibodies from HS patients (100a). With the progression of murine infection, the expression of these competitive idiotypes decreased. Thus, sera from mice infected for 8 weeks inhibited more than sera from mice infected for 16 weeks. Sera from mice infected for 20 to 30 weeks were minimally competitive. Similar data were found when a stimulatory human anti-SEA MAb (E5) (56) was used to raise a specific anti-idiotypic rabbit serum and this serum (after appropriate absorptions) was used to compare idiotypic competition between the homologous human E5 and the idiotypes expressed in mouse serum (32, 100a). Parallels were also found in the ability of both mouse and human anti-SEA antibodies to stimulate either mouse spleen cells or human PBMC. Anti-SEA antibody preparations from INT patients (but not HS patients) were able to stimulate the proliferation of spleen cells from mice infected for 16 weeks (100a), and some mouse anti-SEA MAbs stimulated responses of PBMC from some patients (31). Apparently, mice and men infected with *S. mansoni* produce anti-SEA antibodies that share the expression of some major CRIs.

Chagas' disease

Correlations of idiotypic–anti-idiotypic systems with clinical forms in humans. As described above, auto-anti-idiotypic humoral (132) and cellular (50, 51) responses have been found in patients with Chagas' disease. In both of these situations the anti-idiotypic interactions studied correlated with the clinical form of a patient's disease. The chronic, severe form of infection by *T. cruzi* develops in 20 to 30% of those infected and often involves cardiomyopathy, nerve damage, and enlargement of the heart (chagasic cardiac disease). In a smaller percentage of cases, severe disease is characterized by the digestive forms known as megaesophagus and megacolon. The cardiac lesions are characterized by mononuclear-cell infiltration, consisting of monocytes, lymphocytes, plasma cells, and lymphoblasts, and by fibrosis. In the chronic phase, parasites are not readily observable in either the blood or the tissues, even within these inflammatory lesions (3, 13). The actual etiology of these life-threatening cardiac and digestive lesions is uncertain (3, 13), and they are considered by many to have an autoimmune basis. A variety of autoantibodies and cell-mediated autoimmune reactivities have been reported during Chagas' disease, but their actual relationships to the pathogenesis are uncertain (70, 82). If the disease conditions are directly due to effects of *T. cruzi* organisms, or immune responses to them, it remains to be shown what toxins or antigens are involved (102) and how the organisms or residual antigens induce the lesions when present in small numbers. Lesion formation could also be due to the induction of autoimmune responses against host antigens by parasite antigens which mimic or cross-react with host antigens. A variety of candidate target antigens have been studied, but the actual autoimmune specificities that may be responsible are still not clearly defined.

The data reviewed above, concerning stimulation of auto-anti-idiotypic T cells by anti-*T. cruzi* antibodies, the mode of presentation of these stimulatory agents, and the correlations of these findings with the clinical forms of the infection (Table 1), raise yet another possibility. The idiotypes expressed by anti-EPI antibodies purified from sera of cardiac chagasic patients (IdC) were inherently more stimulatory for PBMC of patients with either cardiac or indeterminate disease (50). Thus, the sera of patients with clinical disease contained antibodies that expressed the most highly stimulatory idiotypes. Furthermore, PBMC from these cardiac chagasic patients were stimulated by IdC directly, without the need for processing of the idiotype or its presentation in conjunction with MHC antigens (30, 49a). This was not so for PBMC from patients with the asymptomatic (indeterminate) form of Chagas' disease (Table 1). Stimulation of PBMC from these patients required that the idiotypes be both processed and presented with MHC. The idiotypes primarily expressed by anti-EPI antibodies from the sera of these patients (IdI) did not stimulate strong T-cell responses, and regardless of the source of responding cells (cardiac or indeterminate patient), proliferation occurred only if the idiotypes were processed and class II MHC presented (30, 49a). As opposed to the situation in schistosomiasis (Table 2), strong antiparasite cellular responses (anti-EPI in this case) correlated with asymptomatic infections and cells from patients with severe disease expressed highly regulated anti-EPI responses (102a). Most of the regulatory response appears to be due to adherent suppressor cells (102a).

It is possible to use these observations and correlations to construct a scenario in which the pathogenesis of severe human Chagas' disease is due not to anti-*T. cruzi* responses, or even to responses against self-*T. cruzi*-mimicked antigens, but to anti-idiotypic responses. This hypothesis proposes that the development of clinical Chagas' disease is due to cellular autoimmune responses against the idiotypes expressed on some anti-*T. cruzi* antibodies. This would be, in the strictest sense, an autoimmune disease, but not against self antigens mimicked by the organism. In this scheme, patients who express the idiotypes associated with asymptomatic disease (IdI) have anti-IdI T cells. Because IdI requires processing and class II MHC presentation and because the level of stimulation by IdI is generally low, the reactions of these anti-idiotypic T cells would be of little pathologic consequence. However, if a patient progressively expresses the idiotypes associated with severe disease (IdC), these directly and strongly T-cell-stimulatory idiotypes would induce unregulated, localized responses. Hypothetically, residual parasite antigen or the few infecting organisms maintained in cardiac tissue would stimulate localized antibody production (plasma cells are a feature of these lesions) or local antiparasite T-cell responses. If the antibody produced or the T-cell receptor involved expressed IdC, they could, without need of processing and presentation, stimulate specific anti-idiotypic T cells (lymphoblasts are a feature of these lesions). Activated anti-idiotypic T cells could release a pattern of lymphokines, which could lead to lesion formation and tissue destruction. The cause of a progression from the expression of IdI to IdC is completely conjectural, but could be based on a wide variety of determining factors, including differences in parasite strains (102) (due to variable antigens, different tissue predilections, etc.), host immunogenetics, maternal and neonatal influences (see below), and/or the induction or relaxation of immunoregulatory systems. This hypothesis constitutes one way in which the idiotypes and anti-idiotypic responses, which we have observed to correlate closely with pathogenesis, could be responsible for lesion formation in Chagas' disease.

Resistance. Auto-anti-idiotypic reactions have been reported to be related to protective immunity in experimental schistosomiasis (119). The potential for anti-idiotypic autoregulation (negative control) of the resistance mechanism(s) in this system was then further studied (121). Lymphoblasts were generated by responses to soluble cercarial immunogens and used to immunize naïve mice against specific anti-SCI lymphoblast receptors. Mice immunized with these lymphoblasts, but not anti-keyhole limpet hemocyanin- or concanavalin A-stimulated lymphoblasts, developed reduced levels of partial resistance upon subsequent immunization with irradiated cercariae. The lowered resistance, induced by the anti-SCI lymphoblasts, to challenge infection was paralleled by lower anti-SCI antibody responses, lower anti-idiotypic responses to a protective anti-SCI MAb, lower SCI-induced proliferative responses, and lower anti-SCI delayed-type hypersensitivity reactions. SCI-specific lymphoblast-immunized mice developed $Lyt1^- L3T4^- Lyt2^+$ anti-lymphoblast T cells specific for the immunizing lymphoblasts. The reduced levels of resistance in mice with these anti-idiotypic cells were hypothesized to be due to their interference with the development and expression of the effector mechanisms responsible for protective immunity in schistosomiasis.

It is also possible that auto-anti-idiotypic, paratope-specific Ab_2 antibodies against protective Ab_1s could contribute to autoimmunizations (positive regulation) and thus to the maintenance of resistance against challenge infections.

Idiotype–anti-idiotype networks related to protective immune responses, either those due to exogenous idiotypic immunization (113, 121) or those seen spontaneously during infection (119), demonstrate the possibility that idiotypes could influence the ability of an infected host to resist further challenge infections or a naïve host to mount an effective form of resistance upon vaccination. Although the interpretations of these observations remain conjectural, one reason to continue to focus on their potential effects concerns vaccine development and usage. The development of vaccines against parasitic infections is one of the major areas of interest in regard to the immunology of these organisms (96, 97, 135, 136). The eventual use of the vaccines to be developed will obviously benefit those entering areas endemic for these diseases (such as immigrants, soldiers, anthropologists, missionaries, engineers, and tropical disease workers). However within endemic areas, the use of these vaccines will encounter logistical difficulties, such as use of a "cold chain," low-temperature maintenance system, distribution systems, and patient control for follow-up boosters. In the context of the current topic, these problems could be compounded by the immunology of the infected population (or their children) within these endemic areas. Immunizations of endemic populations, either already infected or born of infected mothers, may need to take into account the fact that these people might already have anti-idiotypic responses that could aid, or prevent, the induction of protective immunity with a given vaccine (64; see below).

MATERNAL, FETAL, AND PERINATAL INFLUENCES THROUGH IDIOTYPIC INTERACTION

Opportunities

The epidemiology of chronic, endemic parasitic infections often involves situations in which 70% or more of the people living within an endemic area become infected during their childhood and retain their infections (or continually lose and reacquire them) for decades. Because the host and parasite usually maintain a balanced interaction, the majority of the infected populations remain generally fit, but actively infected. It follows that in endemic areas with high prevalence rates for a given infection, infections must occur in many of the pregnant women in the area. Because of the endemic nature of the situation, it follows that most people who are going to become infected are born of mothers who are infected (30, 32, 136). This is in contrast to infections by many other organisms, in which infection occurs, peaks, and either is resolved or kills the patient. In that situation, active infections during pregnancy would be an occasional occurrence. In chronic, endemic infections, as epitomized by many parasitic infections, continuing infections through multiple pregnancies is the rule.

Active infections during pregnancy offer a strong opportunity for both the mother's infection, and her response to the infection, to influence the immune repertoire and capabilities of her children. This goes far beyond the obvious potential for passive immunization during the neonatal period by transfer of the mother's antibodies against the organism in question. Opportunities exist for intrauterine and neonatal exposure to parasite antigens, to the idiotypes on the maternal antibodies against parasite antigens, and to the anti-idiotypic antibodies to those Ab_1, antiparasite antibodies. This could occur through both placental transfer and breast-feeding. These antigens, idiotypes, and anti-idiotypes could either sensitize the offspring to responsiveness (protective or pathogenic) or induce regulatory responses that would be in place in the repertoire of the child, to be anamnestically revived upon later infection.

It is clinically well known that in many parasitic diseases, members of endemic populations often develop less severe disease than do those from outside the endemic areas, such as immigrants or tourists (54, 103, 110, 150). There could be many explanations for such observations. One possibility could be the preexistence of anti-idiotypic, regulatory responses in children born of mothers with appropriate idiotypes that stimulate the control of immunopathogenic responses. These regulatory mechanisms could then come into play more rapidly during the initial encounter of the children with the infection. People born in nonendemic areas would not have been primed in utero for such regulatory responses by antiparasite idiotypes, because their mothers would not have had significant circulating levels of antibodies expressing immunoregulatory idiotypes.

During gestation and breast-feeding there is a real potential for exposure of the children born to mothers with schistosomiasis to schistosomal circulating antigens (16, 47, 104). Transplacental or milk transfer (133) of circulating parasite antigens could therefore lead to perinatal sensitization against these moieties.

Transplacental transfer to the fetus of maternal antibodies to parasitic antigens has been documented for several human parasitic infections (69, 89, 94, 95). It is assumed that it occurs whenever pregnancies are concurrent with active infections and the mother has sufficient levels of circulating antibodies of the correct isotypes. The expected transplacental transfer of anti-schistosomal IgG antibodies from infected mothers to their children has been well documented (69, 89). The epidemiologic patterns of transmission of schistosomiasis show that children usually do not become infected until 2 to 6 years of age (75). Therefore, because of the decay of maternal antibody (69, 89), with this particular infection it would be rare for these antibodies to act in a passive protection mode. However, there remain major potential roles for the idiotypes expressed on the maternal anti-schistosomal antibodies (or anti-anti-schistosomal antibodies [Ab_2s]) transferred to the fetus and neonate. Certainly, experimental studies of perinatal idiotypic interactions make it clear that such exposures can dramatically affect the resulting repertoire of the offspring (9, 48, 68, 78, 84, 87, 92, 105, 126, 127, 144, 146, 153, 154). The opportunity for extensive perinatal idiotypic influences thus exists in regard to many chronic, endemic parasitic infections.

Occurrence

Studies of cord blood mononuclear cell (CBMC) responses of children born of mothers with schistosomiasis have shown strong specific proliferation to schistosomal antigens by cells from many (but not all) neonates (105a). CBMC from children born of uninfected mothers, in an area nonendemic for schistosomiasis, were unresponsive to these schistosomal antigens. Prenatal sensitization to schistosomal antigens has also been shown in human schistosomiasis by skin tests of newborns (15, 145) and macrophage migration inhibitions (145) and in experimental schistosomiasis by the induction of SEA-specific regulation (61, 90). In experimental (63) and human (152) filariasis, transplacentally induced specific unresponsiveness and specific IgE responsiveness, respectively, have been observed. In murine malaria, induction of a specific regulatory response in the offspring born to infected mothers has been shown to be mediated through the induction by maternal IgG of specific suppressor T cells (65). This suppression was responsible for a failure in the ability to effectively vaccinate the offspring of immune mothers (64), but this inability could be overcome by administration of a second dose of vaccine (66).

We recently reported that children born of mothers with schistosomiasis or Chagas' disease have CBMC that responded to immunoaffinity-purified anti-SEA antibodies or anti-EPI antibodies, respectively (46). CBMC-stimulatory antibody preparations could be derived from the sera of the individual babies' mothers (if the sera were obtained early in pregnancy) or from pooled sera of the appropriate patients. CBMC from children born of uninfected mothers (never infected with either *S. mansoni* or *T. cruzi*) did not respond to either anti-SEA or anti-EPI antibodies. This report is the first we know of that demonstrates in utero human idiotypic sensitization in an auto-occurring condition.

Potential Influence

The ramifications of specific in utero sensitization to parasite antigens and the idiotypes on antiparasite antibodies are unknown. Full elucidation of the impact of these events on future infections of these children, as compared with infections of children born of uninfected mothers, would be either ethically impossible or logistically very difficult. If chemotherapy exists, patients must be offered treatment, and their infections thus cannot be studied longitudinally. Longitudinal studies could be done when appropriate drugs are not available. However, the chronic nature of the infections, and the late-stage onset of clinical symptoms in most of these infections, would necessitate that longitudinal studies be continued for decades. Owing to the usual development of the severe clinical form in only a small proportion of those infected, the long-term studies would need to involve large populations of patients. Retrospective studies of familial lines of infected patients from given endemic areas could assist in determining patterns of clinical-form transmission, and correlation of these patterns with idiotypic patterns might be possible. Also, investigations of idiotypic patterns and clinical disease in families that migrate into endemic areas may allow interesting analyses. Potential familial immunogenetic effects on parasitic infections would also have to be considered (1, 106, 138, 140).

OTHER ANTI-IDIOTYPIC STUDIES IN PARASITIC DISEASES

Platelet Interactions in Onchocerciasis

Onchocerciasis, or river blindness, caused by infections with the nematode *Onchocerca volvulus* has, until the recent use of ivermectin, been treated with diethylcarbamazine (DEC). DEC is known to eliminate circulating microfilariae, but also often results in intense, morbidity-related itching and other adverse reactions. It has been demonstrated recently that the microfilaricidal activity of DEC is mediated by activation of blood platelets and an interaction of these platelets with an *O. volvulus* glycoprotein excretory antigen (FEA) (20). FEA, immunoaffinity purified on an anti-FEA rat IgM MAb, augmented DEC-activated platelet killing of microfilariae in a dose-dependent manner, and the investigators wished to study the FEA-FEA receptor interaction on the platelet. However, owing to the high carbohydrate content of FEA, it was deemed unsuitable for standard ^{125}I labeling and use in binding assays. Therefore, mouse anti-idiotypic (Ab$_2$) antibodies against the rat anti-FEA MAb were raised and selected for their ability to inhibit MAb binding to the target antigen (21). The Ab$_2$ MAbs were able to (i) induce the production of anti-*O. volvulus* antibodies (Ab$_3$) in rats; (ii) lead to a dose-dependent cytotoxicity against parasite larvae by DEC-treated platelets, i.e., mimic the action of the FEA; and (iii) bind to four bands of detergent-solubilized platelet surface proteins that may thus be the platelet-binding sites for FEA.

Immunodetection and Immunodiagnosis Systems

Anti-idiotypic MAbs could be attractive candidates for use as immunodiagnostic targets (98). If such a MAb, specific for an idiotype on an anti-parasite MAb, could competitively inhibit the binding of the antiparasite MAb to the antigen, either direct or indirect assays should be feasible. The two reagents needed, both being MAbs, could be prepared and standardized for assay development. These systems have not yet yielded applicable assays (98). However, for both schistosomiasis (98) and malaria (122), idiotype/anti-idiotype competitive assays have been reported to be useful in the detection of parasite antigens. These systems have demonstrated parasite antigens in crude extracts by their abilities to inhibit radioimmunoassays of idiotype–anti-idiotype binding. This type of inhibition system was used to detect a specific sporozoite antigen in sporozoite extracts, and even in homogenates of salivary glands from malaria-infected mosquitoes (122). A similar radioimmunoassay was used to monitor the isolation of a schistosomal antigen through its purification process (98).

SUMMARY AND CONCLUSIONS

The first demonstration that the theoretical use of Ab_2 immunization against an infectious organism was experimentally feasible was with murine infection by the protozoan *T. rhodesiense* (129). Since that time it has become clear that Ab_2 immunization to other parasites and many other infectious organisms can be achieved (67). Because of the difficulties of growing some life cycle stages of certain parasites and the apparent importance of carbohydrate epitopes on some parasite protective antigens, the usefulness of the anti-idiotypic approach to vaccine production may prove to be of considerable importance in some parasitic diseases (130).

Multiple means and studies have demonstrated auto-anti-idiotypic responses (both humoral and cellular) in both experimental and clinical schistosomiasis (31, 32). In fact, these demonstrations have been surprisingly successful (91, 107, 136). Coupled with the same demonstrations for clinical Chagas' disease (30, 50, 51), they suggest that the relative ease of these observations may be based in some way on something shared by these infections. Although these infectious organisms are very different and have very different effects on their hosts, they share the characteristics of causing chronic, endemic infections. It is proposed that chronicity provides the opportunities that allow the development of extensive idiotypic–anti-idiotypic interactions. It is further suggested that the coupling of chronicity with endemicity results in active infections during the gestations that will yield most of the future patients. This situation provides the setting for idiotypic–anti-idiotypic interactions to exert their maximal influence on subsequent immunological expressions. If these suppositions are valid, it would follow that any chronic conditions (infectious, autoimmune, or neoplastic diseases or allogeneic transplants) involving immune responses which are expressed during the child-bearing years should be fertile ground for idiotype/anti-idiotype studies.

The regulatory impact of idiotypic–anti-idiotypic responses is apparent in many experimental immunological systems (9). The validity of this concept for immune responses in everyday clinical settings may turn out to be most easily seen in some parasitic infections. For both schistosomiasis mansoni and schistosomiasis japonicum, the evidence for idiotypic regulation of the immunopathogenicity in experimentally infected mice is extensive (32, 85). Furthermore, for clinical schistosomiasis mansoni, data are accumulating which indicate that these interactions correlate with (101) and, in vitro, can result in (39) immunoregulatory functions.

There are also strong correlations between idiotype–anti-idiotype relationships and clinical disease states in human Chagas' disease (30, 49a). However, it is theorized in this review that in this situation, anti-idiotypic responses may actually contribute to the immunopathology observed.

Opportunities exist for idiotypic–anti-idiotypic responses to play significant roles in the expression of protective immune responses against infections by parasitic organisms. This is true whether the interactions are generated during chronic infections or initiated in utero by being born of an infected mother. It is clear that CBMC of babies born of mothers with either schistosomiasis or Chagas' disease express the ability to respond to idiotypes on their mother's antiparasite antibodies (46). It is possible that such information should be considered for the successful development of antiparasite vaccines designed to be used in endemic populations.

The demonstration that idiotypic–anti-idiotypic responses occur in parasitic diseases, as well as the correlations of such interactions with various clinical forms of these types of infections, opens the door for studies of potential causal relationships between the events. Unfortunately, many of the studies one might plan will be logistically difficult and/or unethical. Nonetheless, some carefully selected studies should be both permissible and useful, and some generally uncommon epidemiologic considerations, such as immigrant populations, may allow unexpected opportunities for study. Also, some experimental animal model systems will probably prove useful in these dissections of cause and effect. However, the model systems may prove difficult if they require the added problems and expense of dealing with the long-term studies needed to work with the offspring of hosts with chronic infections. It is hoped that further information generated through the study of these infections will not only add to the elucidation of their basis, but also contribute to a more complete fundamental understanding of the interactions integral to the internal and external facets of the immune system.

Acknowledgments. It is a pleasure to acknowledge the debt of gratitude owed to the many collaborators with whom I have participated in the studies which have contributed to this chapter. Foremost of these is Giovanni Gazzinelli. Others are T. Abe, Z. Brener, R. Correa-Oliveira, B. Doughty, S. Eloi-Santos, R. Gazzinelli, A. Goes, M. Lima, A. Montesano, E. Novato-Silva, J. Parra, and M. Powell. The coordinating and logistical skills of Judith O'Connell and George L. Freeman, Jr., are also gratefully acknowledged, as is the encouragement of D. Sacks and S. Stewart.

The support of the United Nations Developmental Program/World Bank/World Health Organization Special Programme for Research and Training in Tropical Diseases, the National Institute of Allergy and Infectious Diseases, the Veterans Administration, FINEP and CNPq (Brasilia, Brazil), and the International Development Research Centre (Ottawa, Canada) is also gratefully acknowledged.

REFERENCES

1. **Abdel-Salam, E., A. Abdel-Khalik, A. Abdel-Meguid, W. Barakat, and A. A. F. Mahmoud.** 1986. Association of HLA class I antigens (A1, B5, and CW2) with disease manifestations and infection in human schistosomiasis mansoni in Egypt. *Tissue Antigens* **27**:142–146.

2. **Abe, T., and D. G. Colley.** 1984. Modulation of *Schistosoma mansoni* egg-induced granuloma formation. III. Evidence for an anti-idiotypic, I-J-positive, I-J-restricted, soluble T suppressor factor. *J. Immunol.* **132**:2084–2088.

3. **Andrade, Z.** 1983. Mechanisms of myocardial damage in *Trypanosoma cruzi* infection. *CIBA Found. Symp.* **99**:214–233.

4. **Andrade, Z., and K. S. Warren.** 1964. Mild prolonged schistosomiasis in mice: alterations in host response with time and the development of portal fibrosis. *Trans. R. Soc. Trop Med. Hyg.* **58**:53–57.

5. **Asherson, G. L., V. Colizzi, and M. Zembala.** 1986. An overview of T-suppressor cell circuits. *Annu. Rev. Immunol.* **4**:37–68.

6. **Bentley, A. G., S. M. Phillips, R. J. Kaner, V. J. Theodorides, G. P. Linette, and B. L. Doughty.** 1985. In vitro delayed hypersensitivity granuloma formation: development of an antigen-coated bead model. *J. Immunol.* **134**:4163–4169.

7. **Bhogal, B. S., K. H. Nollstadt, Y. D. Karkhanis, D. M. Schmatz, and E. B. Jacobson.** 1988. Anti-idiotypic antibody with potential use as an *Eimeria tenella* sporozoite antigen surrogate for vaccination of chickens against coccidiosis. *Infect. Immun.* **56**:1113–1119.

8. **Bogen, B., B. Malissen, and W. Haas.** 1986. Idiotype-specific T cell clones that recognize syngeneic immunoglobulin fragments in the context of class II molecules. *Eur. J. Immunol.* **16**:1373–1378.

9. **Bona, C. A.** 1987. *Idiotypes and Lymphocytes.* Academic Press, Inc., New York.

10. **Boros, D. L.** 1986. Immunoregulation of granuloma formation in murine schistosomiasis mansoni. *Ann. N.Y. Acad. Sci.* **465**:313–323.

11. **Bottomly, K., and P. H. Mauer.** 1980. Antigen-specific helper T cells required for dominant production of an idiotype (ThId) are not under immune response (Ir) gene control. *J. Exp. Med.* **152**:1571–1582.

12. **Bottomly, K., and D. E. Moiser.** 1981. Antigen-specific helper T cells required for dominant idiotype expression are not H-2 restricted. *J. Exp. Med.* **154**:411–421.

13. **Brener, Z.** 1987. Pathogenesis and immunopathology of chronic Chagas' disease. *Mem. Inst. Oswaldo Cruz Rio J.* **82(Suppl.)**:205–213.

14. **Brown, J. C., and L. S. Rodkey.** 1979. Autoregulation of an antibody response via network-induced auto-antiidiotype. *J. Exp. Med.* **150**:67–85.

15. **Camus, D., Y. Carlier, J. C. Bina, R. Borojevic, A. Prata, and A. Capron.** 1976. Sensitization to *Schistosoma mansoni* antigen in uninfected children born to infected mothers. *J. Infect. Dis.* **134:**405–408.

16. **Carlier, Y., H. Nzeyimana, D. Bout, and A. Capron.** 1980. Evaluation of circulating antigens by a sandwich radioimmunoassay, and of antibodies and immune complexes, in *Schistosoma mansoni*-infected African parturients and their newborn children. *Am. J. Trop. Med. Hyg.* **29:**74–81.

17. **Cerny, J., and M. J. Caulfield.** 1981. Stimulation of specific antibody-forming cells in antigen-primed nude mice by the adoptive transfer of syngeneic anti-idiotypic T cells. *J. Immunol.* **126:**2262–2266.

18. **Cerny, J., R. Cronkhite, and J. T. Stout.** 1986. Rapid changes in the regulatory potential of autologous anti-idiotopic T cells during an antigen-driven primary response. *J. Immunol.* **136:**3597–3606.

19. **Cerny, J., J. S. Smith, C. Webb, and P. W. Tucker.** 1988. Properties of anti-idiotypic T cell lines propagated with syngeneic B lymphocytes. I. T cells bind intact idiotypes and descriminate between the somatic idiotypic variants in a manner similar to the anti-idiotopic antibodies. *J. Immunol.* **141:**3718–3725.

20. **Cesbron, J.-Y., A. Capron, B. B. Vargaftif, M. Legarde, J. Pincemail, P. Braquet, H. Taelman, and M. Joseph.** 1987. Platelets mediate the action of diethylcarbamazine on microfilariae. *Nature* (London) **325:**533–536.

21. **Cesbron, J.-Y., M. Hayasaki, M. Joseph, C. Lutsch, J.-M. Grzych, and A. Capron.** 1988. *Onchocerca volvulus.* Monoclonal anti-idiotype antibody as antigen signal for the microfilaricidal cytotoxicity of diethylcarbamazine-treated platelets. *J. Immunol.* **141:**279–285.

22. **Chensue, S. W., D. L. Boros, and C. S. David.** 1983. Regulation of granulomatous inflammation in murine schistosomiasis. II. T suppressor cell-derived, I-C subregion-encoded soluble suppressor factor mediates regulation of lymphokine production. *J. Exp. Med.* **157:**219–230.

23. **Colley, D. G.** 1975. Immune responses to a soluble schistosomal egg antigen preparation during chronic primary infections with *Schistosoma mansoni. J. Immunol.* **115:**150–156.

24. **Colley, D. G.** 1976. Adoptive suppression of granuloma formation. *J. Exp. Med.* **143:**696–700.

25. **Colley, D. G.** 1981. Immune responses and immunoregulation in experimental and clinical schistosomiasis, p. 1–81. *In* J. M. Mansfield (ed.), *Parasitic Diseases,* vol. 1. *The Immunology.* Marcel Dekker, Inc., New York.

26. **Colley, D. G.** 1981. T lymphocytes which contribute to the immunoregulation of granuloma formation in chronic murine schistosomiasis. *J. Immunol.* **126:**1465–1468.

27. **Colley, D. G.** 1987. Dynamics of the human immune response to schistosomes, p. 315–332. *In* A. A. F. Mahmoud (ed.), *Bailliere's Clinical Tropical Medicine and Communicable Diseases,* vol. 2, no. 2. *Schistosomiasis.* The W. B. Saunders Co., Ltd., Eastbourne, England.

28. **Colley, D. G., and M. D. Colley.** 1989. Protective immunity and vaccines to schistosomiasis. *Parasitol. Today* **5:**350–354.

29. **Colley, D. G., A. S. Garcia, J. R. Lambertucci, J. C. Parra, N. Katz, R. S. Rocha, and G. Gazzinelli.** 1986. Immune responses during human schistosomiasis. XII. Differential responsiveness in patients with hepatosplenic disease. *Am. J. Trop. Med. Hyg.* **35:**793–802.

30. **Colley, D. G., R. T. Gazzinelli, Z. Brener, M. A. Montesano, S. M. Eloi-Santos, R. Correa-Oliveira, J. C. Parra, M. S. Lima, and G. Gazzinelli.** 1990. Idiotypes and anti-idiotypic T lymphocytes in chronic endemic parasitic diseases. *In* F. UytdeHaag, J. Urbain, and A. Osterhaus (ed.), *Idiotypes in Biology and Medicine,* Elsevier Biomedical Press, Amsterdam, in press.

31. **Colley, D. G., A. M. Goes, B. L. Doughty, R. Correa-Oliveira, J. C. Parra, K. M. Lairmore, M. A. Montesano, R. S. Rocha, and G. Gazzinelli.** 1989. Antiidiotypic T cells and factors in murine and human schistosomiasis, p. 367–378. *In* J. G. Kaplan, D. R. Green, and R. C. Bleackley (ed.), *The Cellular Basis of Immune Modulation.* Alan R. Liss, Inc., New York.

32. **Colley, D. G., M. A. Montesano, S. M. Eloi-Santos, M. R. Powell, R. Correa-Oliveira, R. S. Rocha, and G. Gazzinelli.** 1989. Idiotype networks in schistosomiasis, p. 179–190. *In* K. P. W. J. McAdam (ed.), *New Strategies in Parasitology.* Churchill Livingstone, Edinburgh.

33. **Colley, D. G., J. C. Parra, M. A. Montesano, M. Lima, E. Nascimento, B. L. Doughty, A. Goes, and G. Gazzinelli.** 1987. Immunoregulation in human schistosomiasis by idiotypic interactions and lymphokine-mediated mechanisms. *Mem. Inst. Oswaldo Cruz Rio J.* **82(Suppl.):**105–109.

34. **Cosenza, H.** 1976. Detection of anti-idiotype reactive cells in response to phosphorylcholine. *Eur. J. Immunol.* **6:**114–116.

35. **Dissous, C., J.-M. Grzych, and A. Capron.** 1982. *Schistosoma mansoni* surface antigen defined by a rat monoclonal IgG2a. *J. Immunol.* **129:**2232–2234.

36. **Dissous, C., A. Prata, and A. Capron.** 1984. Human antibody response to *Schistosoma mansoni* surface antigens defined by protective monoclonal antibodies. *J. Infect. Dis.* **149:**227–233.

37. **Donelson, J.** 1988. Unsolved mysteries of trypanosome antigenic variation, p. 371–400. *In* P. T. Englund and A. Sher (ed.), *The Biology of Parasitism.* Alan R. Liss, Inc., New York.

38. **Dorf, M. E., and B. Benacerraf.** 1984. Suppressor cells and immunoregulation. *Annu. Rev. Immunol.* **2:**127–157.

39. **Doughty, B. L., A. M. Goes, J. C. Parra, R. Rocha, J. C. Cone, D. G. Colley, and G. Gazzinelli.** 1989. Anti-idiotypic T cells in human schistosomiasis. *Immunol. Invest.* **18:**373–388.

40. **Doughty, B. L., A. M. Goes, J. C. Parra, R. S. Rocha, N. Katz, D. G. Colley, and G. Gazzinelli.** 1987. Granulomatous hypersensitivity to *Schistosoma mansoni* egg antigens in human schistosomiasis. I. Granuloma formation and modulation around polyacrylamide antigen-conjugated beads. *Mem. Inst. Oswaldo Cruz Rio J.* **82(Suppl.):**47–54.

41. **Doughty, B. L., E. A. Ottesen, T. E. Nash, and S. M. Phillips.** 1984. Delayed hypersensitivity granuloma formation around schistosomiasis mansoni eggs *in vitro.* III. Granuloma formation and modulation in human schistosomiasis mansoni. *J. Immunol.* **133:**993–997.

42. **Dunn, E. B., and K. Bottomly.** 1985. T15-specific helper T cells: analysis of idiotype specificity by competitive inhibition analysis. *Eur. J. Immunol.* **15:**728–732.

43. **Dunn, E. B., J. Kim, and K. Bottomly.** 1986. A cloned T cell line that selectively augments antibody responses of phosphorylcholine-specific B cells bearing the T15 idiotype. *J. Mol. Cell. Immunol.* **2:**209–217.

44. **Eddy, N. K., R. J. Apple, and J. G. Michael.** 1987. Idiotype regulation of the anti-bovine serum albumin response. I. Generation of an idiotype-specific Lyt-1$^-$2$^+$ suppressor T cell. *J. Immunol.* **138:**1693–1698.

45. **El Naggar, A., and D. G. Colley.** 1982. Modulation of *Schistosoma mansoni* egg-induced granuloma formation. II. Soluble suppressor activity from lymphoid cells during chronic infection. *Cell. Immunol.* **72:**151–156.

46. **Eloi-Santos, S. M., E. Novato-Silva, V. M. Maselli, G. Gazzinelli, D. G. Colley, and R. Correa-Oliveira.** 1989. Idiotypic sensitization in utero of children born to mothers with schistosomiasis or Chagas' disease. *J. Clin. Invest.* **84:**1028–1031.

47. **Feldmeier, H., J. A. Nogueira-Queiroz, M. A. Peixoto-Queiroz, E. Doehring, J. P. Dessaint, J. E. de Alencar, A. A. Dafalla, and A. Capron.** 1986. Detection and quantification of circulating antigen in schistosomiasis by monoclonal antibody II. The quantification of circulating antigens in human schistosomiasis mansoni and haematobium: relationship to intensity of infection and disease status. *Clin. Exp. Immunol.* **65:**232–243.

48. **Freitas, A. A., O. Burlen, A. Coutinho.** 1988. Selection of antibody repertoires by anti-idiotypes can occur at multiple steps of B cell differentiation. *J. Immunol.* **140:**4097–4102.

49. **Garb, K. S., A. B. Stavitsky, G. R. Olds, J. W. Tracy, and A. A. F. Mahmoud.** 1982. Immune regulation in murine schistosomiasis japonica. Inhibition of in vitro antigen- and mitogen-induced cellular responses by splenocyte culture supernates and by purified fractions from serum of chronically infected mice. *J. Immunol.* **129:**2752–2758.

49a.**Gazzinelli, R. T., G. Gazzinelli, J. R. Cançado, J. E. Cardoso, Z. Brener, and D. G. Colley.** 1990. Two modes of idiotypic stimulation of T lymphocytes from patients with Chagas' disease: correlations with clinical forms of infection. Submitted for publication.

50. **Gazzinelli, R. T., M. J. F. Morato, R. M. B. Nunes, J. R. Cancado, Z. Brener, and G. Gazzinelli.** 1988. Idiotype stimulation of T lymphocytes from *Trypanosoma cruzi*-infected patients. *J. Immunol.* **140:**3167–3172.

51. **Gazzinelli, R. T., J. R. C. Parra, R. Correa-Oliveira, R. J. Cancado, R. S. Rocha, G. Gazzinelli, and D. G. Colley.** 1988. Idiotypic/anti-idiotypic interactions in schistosomiasis and/or Chagas' disease. *Am. J. Trop. Med. Hyg.* **39:**288–294.

52. **Geha, R. S.** 1983. Presence of circulating anti-idiotype-bearing cells after booster immunization with tetanus toxoid (TT) and inhibition of anti-TT antibody synthesis by auto-anti-idiotypic antibody. *J. Immunol.* **130:**1634–1639.

53. **Geha, R. S.** 1986. Idiotypic interactions in the treatment of human diseases. *Adv. Immunol.* **39:**255–297.

54. **Gelfand, M.** 1967. *A Clinical Study of Intestinal Bilharziasis (Schistosoma mansoni) in Africa.* Edward Arnold, London.

55. **Ghosh, S. K., J. Wong, and R. B. Bankert.** 1987. Idiotype-specific T lymphocytes responsible for the selection of somatic variants of a B cell hybrid. *J. Immunol.* **138:**2230–2235.

56. **Goes, A., R. S. Rocha, G. Gazzinelli, and B. L. Doughty.** 1989. Production and characterization of human monoclonal antibodies against *Schistosoma mansoni. Parasite Immunol.* **11:**695–711.

57. **Goidl, E. A., A. F. Schrater, G. W. Siskind, and G. J. Thorbecke.** 1979. Production of auto-anti-idiotypic antibody during the normal immune response to TNP-Ficoll. II. Hapten-reversible inhibition of anti-TNP plaque-forming cells by immune serum as an assay for auto-anti-idiotypic antibody. *J. Exp. Med.* **150:**154–165.

58. **Grzych, J.-M., M. Capron, H. Bazin, and A. Capron.** 1982. In vitro and in vivo effector function of rat IgG2a monoclonal anti-*S. mansoni* antibodies. *J. Immunol.* **129:**2739–2743.

59. **Grzych, J.-M., M. Capron, C. Dissous, and A. Capron.** 1984. Blocking activity of rat monoclonal antibodies in experimental schistosomiasis. *J. Immunol.* **133:**998–1004.

60. **Grzych, J. M., M. Capron, P. H. Lambert, C. Dissous, S. Torres, and A. Capron.** 1985. An anti-idiotype vaccine against experimental schistosomiasis. *Nature* (London) **316:**74–76.

61. **Hang, L. M., D. L. Boros, and K. S. Warren.** 1974. Induction of immunological hyporesponsiveness to granulomatous hypersensitivity in *Schistosoma mansoni* infection. *J. Infect. Dis.* **130:**515–522.

62. **Hannestad, K., G. Kristoffersen, and J. P. Briand.** 1986. The T lymphocyte response to syngeneic λ2 light chain idiotypes: significance of individual amino acids revealed by variant λ2 chains and idiotope-mimicking chemically synthesized peptides. *Eur. J. Immunol.* **16:**889–893.

63. **Haque, A., and A. Capron.** 1982. Transplacental transfer of rodent microfilariae induces antigen-specific tolerance in rats. *Nature* (London) **299**:361–363.

64. **Harte, P. G., J. B. De Souza, and J. H. L. Playfair.** 1982. Failure of malaria vaccination in mice born to immune mothers. *Clin. Exp. Immunol.* **49**:509–516.

65. **Harte, P. G., and J. H. L. Playfair.** 1983. Failure of malaria vaccination in mice born to immune mothers. II. Induction of specific suppressor cells by maternal IgG. *Clin. Exp. Immunol.* **51**:157–164.

66. **Harte, P. G., N. Rogers, G. A. T. Targett, and J. H. L. Playfair.** 1984. Maternal inhibition of malaria vaccination in mice can be overcome by giving a second dose of vaccine. *Immunology* **53**:401–409.

67. **Hiernaux, J. R.** 1988. Idiotypic vaccines and infectious diseases. *Infect. Immun.* **56**:1407–1413.

68. **Hiernaux, J. R., C. Bona, and P. J. Baker.** 1981. Neonatal treatment with low doses of anti-idiotypic antibody leads to the expression of a silent clone. *J. Exp. Med.* **153**:1004–1008.

69. **Hillyer, G. V., R. Menendez-Corrada, R. Lluberes, and F. Hernandez-Morales.** 1970. Evidence of transplacental passage of specific antibody in schistosomiasis mansoni in man. *Am. J. Trop. Med. Hyg.* **19**:289–291.

70. **Hudson, L., and P. J. Hindmarsh.** 1985. The relationship between autoimmunity and Chagas' disease: causal or coincidental? *Curr. Top. Microbiol. Immunol.* **117**:167–177.

71. **Janeway, C.** 1986. Varieties of idiotype-specific helper T cells: a commentary. *J. Mol. Cell Immunol.* **2**:265–267.

72. **Janeway, C. A., Jr., J. P. Tite, K. Bottomly, and P. Flood.** 1985. The role of immunoglobulin and B cells in T cell ontogeny and repertoire formation: a commentary. *J. Mol. Cell. Immunol.* **2**:118–120.

73. **Jayaraman, S., and C. J. Bellone.** 1982. Hapten-specific responses to the phenyltrimethylamino hapten. II. Regulation of delayed-type hypersensitivity by genetically restricted, anti-idiotypic suppressor T cells induced by the monvalent antigen L-tyrosine-p-azophenyltrimethylammonium. *Eur. J. Immunol.* **12**:278–284.

74. **Jayaraman, S., J. E. Swierkosz, and C. J. Bellone.** 1982. T cell replacing factor substitutes for an I-J+ idiotype-specific T helper cell. *J. Exp. Med.* **155**:641–646.

75. **Jordan P.** 1985. *Schistosomiasis, The St Lucia project.* Cambridge University Press, Cambridge.

76. **Jorgensen, R., and K. Hannestad.** 1982. Helper T cell recognition of the variable domains of a mouse myeloma protein (315). *J. Exp. Med.* **155**:1587–1596.

77. **Kawahara, D. J., P. Marrack, and J. W. Kappler.** 1986. H-2-linked Ir gene control of VH determinant(s)-specific helper T cells. *J. Mol. Cell. Immunol.* **2**:255–264.

78. **Kearney, J. F., and M. Vakil.** 1986. Idiotype-directed interactions during ontogeny play a major role in the establishment of the adult B cell repertoire. *Immunol. Rev.* **94**:39–50.

79. **Kelsoe, G., and J. Cerny.** 1979. Reciprocal expansion of idiotypic and anti-idiotypic clones following antigen stimulation. *Nature* (London) **279**:333–334.

80. **Kelsoe, G., D. Isaak, and J. Cerny.** 1980. Thymic requirement for cyclical idiotypic and reciprocal anti-idiotypic immune response to a T-independent antigen. *J. Exp. Med.* **151**:289–300.

81. **Khalife, J., M. Capron, A. Capron, J.-M. Grzych, A. E. Butterworth, D. W. Dunne, and J. H. Ouma.** 1986. Immunity in human schistosomiasis mansoni: regulation of protective immune mechanisms by IgM blocking antibodies. *J. Exp. Med.* **164**:1626–1640.

82. **Kierszenbaum, F.** 1986. Autoimmunity in Chagas' disease. *J. Parasitol.* **72**:201–211.

83. **King, C. H., R. R. Lett, J. Nanduri, S. El Ibiary, P. A. S. Peters, G. R. Olds, and A. A. F. Mahmoud.** 1987. Isolation and characterization of a protective antigen for adjuvant-free immunization against *Schistosoma mansoni. J. Immunol.* **139**:4218–4224.

84. **Kohler, H., J. Urbain, and P. A. Cazenave.** 1984. *Idiotypy in Biology and Medicine.* Academic Press, Inc., New York.

85. **Kresina, T. F., and G. R. Olds.** 1986. Concomitant cellular and humoral expression of a regulatory cross-reactive idiotype in acute *Schistosoma japonicum* infection. *Infect. Immun.* **53**:90–94.

86. **Kresina, T. F., and G. R. Olds.** 1989. Antiidiotypic antibody vaccine in murine schistsomiasis mansoni comprising the internal image of antigen. *J. Clin. Invest.* **83**:912–920.

87. **Kresina, T. F., and A. Nisonoff.** 1983. Passive transfer of the idiotypically suppressed state by serum from suppressed mice and transfer of suppression from mothers to offspring. *J. Exp. Med.* **157**:15–23.

88. **Lambert, P. H., M. Goldman, L. M. Rose, and P. A. Morel.** 1982. Idiotypic interactions and immunopathology. *Ann. Inst. Pasteur Immunol.* **133C**:239–243.

89. **Lees, R. E. M., and P. Jordan.** 1968. Transplacental transfer of antibodies to *Schistosoma mansoni* and their persistence in infants. *Trans. R. Soc. Trop. Med. Hyg.* **62**:630–631.

90. **Lewert, R. M., and S. Mandlowitz.** 1969. Schistosomiasis: prenatal induction of tolerance to antigens. *Nature* (London) **224**:1029–1030.

91. **Lima, M. S., G. Gazzinelli, E. Nacimento, J. Cavalho-Parra, M. A. Montesano, and D. G. Colley.** 1986. Immune responses during human schistosomiasis mansoni: evidence for antiidiotypic T lymphocyte responsiveness. *J. Clin. Invest.* **78**:983–988.

92. **Martinez, A. C., M. L. Toribio, A. De La Hera, P.-A. Cazenave, and A. Coutinho.** 1986. Maternal transmission of idiotypic network interactions selecting available T cell repertoires. *Eur. J. Immunol.* **16**:1445–1447.

93. **Mathew, R. C., and D. L. Boros.** 1986. Regulation of granulomatous inflammation in murine schistosomiasis. III. Recruitment of antigen-specific I-J+ T suppressor cells of the granulomatous response by an I-J+ soluble suppressor factor. *J. Immunol.* **136**:1093–1099.

94. **McGregor, I. A.** 1972. Immunology of malarial infection and its possible consequences. *Br. Med. Bull.* **28**:22–27.

95. **Miles, M. A., V. Macedo, C. Castro, and C. C. Draper.** 1975. *Trypanosoma cruzi*—prenatal transfer of maternal antibody in man. *Trans. R. Soc. Trop. Med. Hyg.* **69**:286.

96. **Mitchell, G. F.** 1989. Problems specific to parasite vaccines. *Parasitology* **98**:S19–S28.

97. **Mitchell, G. F.** 1989. An update on candidate malaria vaccines. *Parasitology* **98**:S29–S47.

98. **Mitchell, G. F., E. G. Garcia, and K. M. Cruise.** 1983. Competitive radioimmunoassays using hybridoma and anti-idiotype antibodies in indentification of antibody responses to, and antigens of, *Schistosoma japonicum. Aust. J. Exp. Biol. Med.* **61**:27–36.

99. **Mitchell, G. F., E. G. Garcia, K. M. Cruise, W. U. Tiu, and R. E. Hocking.** 1982. Lung granulomatous hypersensitivity to eggs of *Schistosoma japonicum* in mice analysed by a radioisotopic assay and effects of hybridoma (idiotype) sensitization. *Aust. J. Exp. Biol. Med. Sci.* **60**:401–416.

100. **Molineaux, L., R. Cornille-Brogger, H. M. Mathous, and J. Storey.** 1978. Longitudinal serological study of malaria in infants exposed to, or protected from, transmission from birth. *Bull. W.H.O.* **56**:573–578.

100a. **Montesano, M. A., G. L. Freeman, Jr., G. Gazzinelli, and D. G. Colley.** 1990. Expression of cross-reactive, shared idiotypes on anti-SEA antibodies from humans and mice with schistosomiasis. Submitted for publication.

101. **Montesano, M. A., M. S. Lima, R. Correa-Oliveira, G. Gazzinelli, and D. G. Colley.** 1989. Immune responses during human schistosomiasis mansoni. XVI. Idiotypic differences in antibody preparations from patients with different clinical forms of infection. *J. Immunol.* **142**:2501–2506.

102. **Morato, M. J. F., Z. Brener, J. R. Cancado, R. M. B. Nunes, E. Chiari, and G. Gazzinelli.**

1986. Cellular immune responses of chagasic patients to antigens derived from different strains and clones. *Am. J. Trop. Med. Hyg.* **35**:505–511.

102a. **Morato, M. J. F., J. R. Cançado, D. G. Colley, Z. Brener, and G. Gazzinelli.** 1990. B lymphocyte presentation of epimastigote antigens to T lymphocytes in human Chagas' disease. Submitted for publication.

103. **Nash, T. E., A. W. Cheever, E. O. Ottesen, and J. A. Cook.** 1982. Schistosome infections in humans: perspectives and recent findings. *Ann. Intern. Med.* **97**:740–754.

104. **Nash, T. E., and A. M. Deelder.** 1985. Comparison of four schistosome excretory-secretory antigens: phenol sulfuric test active peak, cathodic circulating antigen, gut-associated proteoglycan, and circulating anodic antigen. *Am. J. Trop. Med. Hyg.* **34**:236–241.

105. **Nisonoff, A., M. F. Gurish, and T. F. Kresina.** 1983. Maternal-fetal transfer of a state of idiotypic suppression. *Ann. N.Y. Acad. Sci.* **418**:40–47.

105a. **Novato-Silva, E., G. Gazzinelli, and D. G. Colley.** 1990. Immune responses during human schistosomiasis mansoni. XVII. Immunologic status of pregnant women and their neonates. Submitted for publication.

106. **Ohta, N., M. Hayashi, L. C. Tormis, B. L. Blas, J. S. Nosenas, and T. Sasazuki.** 1987. Immunogenetic factors involved in the pathogenesis of distinct clinical manifestations of schistosomiasis japonica in the Philippine population. *Trans. R. Soc. Trop. Med. Hyg.* **81**:292–296.

107. **Olds, G. R., and T. F. Kresina.** 1985. Network interactions in *Schistosoma japonicum* infection. Identification and characterization of serologically distinct immunoregulatory auto-antiidiotypic antibody population. *J. Clin. Invest.* **76**:2338–2347.

108. **Olds, G. R., and T. F. Kresina.** 1989. Characterization of a family of monoclonal antibodies which bind *S. japonicum* egg antigens and express an interstrain cross reactive idiotype. Submitted for publication.

109. **Olds, G. R., R. Olveda, J. W. Tracy, and A. A. F. Mahmoud.** 1982. Adoptive transfer of modulation of granuloma formation and hepatosplenic disease in murine schistosomiasis japonica by serum from chronically infected mice. *J. Immunol.* **128**:1391–1393.

110. **Ottesen, E. A.** 1984. Immunological aspects of lymphatic filariasis and onchocerciasis in man. *Trans. R. Soc. Trop. Med. Hyg.* **78**:9–18.

111. **Parra, J. C., M. S. Lima, G. Gazzinelli, and D. G. Colley.** 1988. Immune responses during human schistosomiasis mansoni. XV. Anti-idiotypic T cells can recognize and respond to anti-SEA idiotypes directly. *J. Immunol.* **140**:2401–2405.

112. **Paul, W. E., and C. Bona.** 1982. Regulatory idiotypes and immune networks: a hypothesis. *Immunol. Today* **3**:230–234.

113. **Percy, A. J., and D. A. Harn.** 1988. Monoclonal anti-idiotypic and anti-anti-idiotypic antibodies from mice immunized with a protective monoclonal antibody against *Schistosoma mansoni*. *J. Immunol.* **140**:2760–2762.

114. **Perrin, P. J., and S. M. Phillips.** 1988. The molecular basis of granuloma formation in schistosomiasis. I. A T cell-derived suppressor effector factor. *J. Immunol.* **141**:1714–1719.

115. **Perrin, P. J., and S. M. Phillips.** 1989. The molecular basis of granuloma formation in schistosomiasis. III. In vivo effects of a T cell-derived suppressor effector factor and IL-2 on granuloma formation. *J. Immunol.* **143**:649–654.

116. **Perrin, P. J., M. B. Prystowsky, and S. M. Phillips.** 1989. The molecular basis of granuloma formation in schistosomiasis. II. Analogies of a T cell-derived suppressor effector factor to the T cell receptor. *J. Immunol.* **142**:985–991.

117. **Philipp, M., T. B. Davis, N. Storey, and C. S. Carlow.** 1988. Immunity in filariasis: perspectives for vaccine development. *Annu. Rev. Microbiol.* **42**:685–716.

118. **Phillips, S. M., and D. G. Colley.** 1978. Immunologic aspects of host responses to schis-

tosomiasis: resistance, immunopathology and eosinophil involvement. *Prog. Allergy* **24**:49–182.

119. **Phillips, S. M., E. G. Fox, N. G. Fathelbab, and D. Walker.** 1986. Epitopic and paratopically directed anti-idiotypic factors in the regulation of resistance to murine schistosomiasis mansoni. *J. Immunol.* **137**:2339–2347.

120. **Phillips, S. M., and P. J. Lammie.** 1986. Immunopathology of granuloma formation and fibrosis in schistosomiasis. *Parasitol. Today* **2**:296–302.

121. **Phillips, S. M., P. J. Perrin, D. J. Walker, N. G. Fathelbab, G. P. Linette, and M. A. Idris.** 1988. The regulation of resistance to *Schistosoma mansoni* by auto-anti-idiotypic immunity. *J. Immunol.* **141**:1728–1733.

122. **Potocnjak, P., F. Zavala, R. Nussenzweig, and V. Nussenzweig.** 1982. Inhibition of idiotype-anti-idiotype interaction for detection of a parasite antigen: a new immunoassay. *Science* **215**:1637–1639.

123. **Powell, M. R., and D. G. Colley.** 1985. Demonstration of splenic auto-anti-idiotypic plaque-forming cells in mice infected with *Schistosoma mansoni. J. Immunol.* **134**:4140–4145.

124. **Powell, M. R., and D. G. Colley.** 1987. Anti-idiotypic T lymphocyte responsiveness in murine schistosomiasis mansoni. *Cell. Immunol.* **104**:377–385.

125. **Rose, L. M., M. Goldman, and P.-H. Lambert.** 1982. Simultaneous induction of an idiotype, corresponding anti-idiotypic antibodies, and immune complexes during African trypanosomiasis in mice. *J. Immunol.* **128**:79–85.

126. **Rothstein, T. L., and A. P. Vastola.** 1984. Homologous monoclonal antibodies induce idiotope-specific suppression in neonates through maternal influence and in adults exposed during fetal and neonatal life. *J. Immunol.* **133**:1151–1154.

127. **Rubinstein, L. J., M. Yeh, and C. A. Bona.** 1982. Idiotype-anti-idiotype network. II. Activation of silent clones by treatment at birth with idiotypes is associated with the expansion of idiotype specific helper T cells. *J. Exp. Med.* **153**:506–521.

128. **Sacks, D. L.** 1985. Molecular mimicry of parasitic antigens using anti-idiotypic antibodies. *Curr. Top. Microbiol. Immunol.* **119**:45–55.

129. **Sacks, D. L., K. M. Esser, and A. Sher.** 1982. Immunization of mice against African trypanosomiasis using anti-idiotypic antibodies. *J. Exp. Med.* **155**:1108–1119.

130. **Sacks, D. L., L. V. Kirchoff, S. Hieny, and A. Sher.** 1985. Molecular mimicry of a carbohydrate epitope on a major surface glycoprotein of *Trypanosoma cruzi* by using anti-idiotypic antibodies. *J. Immunol.* **135**:4155–4159.

131. **Sacks, D. L., and A. Sher.** 1983. Evidence that anti-idiotype induced immunity to experimental African trypanosomiasis is genetically restricted and requires recognition of combining site-related idiotopes. *J. Immunol.* **131**:1511–1515.

132. **Sadigursky, M., B. F. Von Kreuter, and C. A. Santos-Buch.** 1988. Development of chagasic autoimmune myocarditis associated with anti-idiotype reaction. *Mem. Inst. Oswaldo Cruz Rio J.* **83(Suppl. I)**:363–366.

133. **Santoro, F., Y. Carlier, R. Borojevic, D. Bout, P. Tachon, and A. Capron.** 1977. Parasite "M" antigen in milk from mothers infected with *Schistosoma mansoni* (preliminary report). *Ann. Trop. Med. Parasitol.* **71**:121–122.

134. **Schrater, A. F., E. A. Goidl, G. J. Thorbecke, and G. W. Siskind.** 1979. Production of auto-anti-idiotypic antibody during the normal response to TNP-Ficoll. I. Occurrence in AKR/J and BALB/c mice of hapten-augmentable, anti-TNP plaque-forming cells and their accelerated appearance in recipients of immune spleen cells. *J. Exp. Med.* **150**:138–153.

135. **Sher, A.** 1988. Vaccination against parasites: special problems imposed by the adaptation of parasitic organisms to the host immune response, p. 169–182. *In* A. Sher and P. Englund (ed.), *The Biology of Parasitism.* Alan R. Liss, Inc., New York.

136. **Sher, A., and D. G. Colley.** 1989. Immunoparasitology, p. 957–983. *In* W. E. Paul (ed.), *Fundamental Immunology,* 2nd ed. Raven Press, New York.

137. **Sher, A., S. L. James, R. Correa-Oliveira, S. Hieny, and E. Pearce.** 1989. Schistosome vaccines: current progress and future prospects. *Parasitology* **98:**S61–S68.

138. **Sher, A., and P. A. Scott.** 1982. Genetic factors influencing the interaction of parasites with the immune system. *Clin. Immunol. Allergy* **2:**489–510.

139. **Sidner, R. A., C. E. Carter, and D. G. Colley.** 1987. Modulation of *Schistosoma japonicum* pulmonary egg granulomas with monoclonal antibodies. *Am. J. Trop. Med. Hyg.* **36:**361–370.

140. **Skamene, E.** 1985. *Genetic Control of Host Resistance to Infection and Malignancy.* Alan R. Liss, Inc., New York.

141. **Snary, D.** 1983. Cell surface glycoproteins of *Trypanosoma cruzi:* protective immunity in mice and antibody levels in human chagasic sera. *Trans. R. Soc. Trop. Med. Hyg.* **77:**126–129.

142. **Snary, D., M. A. J. Ferguson, M. T. Scott, and A. K. Allen.** 1981. Cell surface antigens of *Trypanosoma cruzi:* use of monoclonal antibodies to identify and isolate an epimastigote specific glycoprotein. *Mol. Biochem. Parasitol.* **3:**343–356.

143. **Stavitsky, A. B.** 1987. Immune regulation in schistosomiasis japonica. *Immunol. Today* **8:**228–233.

144. **Stein, K. E., and T. Soderstrom.** 1984. Neonatal administration of idiotype or antiidiotype primes for protection against *Escherichia coli* K13 infection in mice. *J. Exp. Med.* **160:**1001–1011.

145. **Tachon, P., and R. Borojevic.** 1978. Mother-child relation in human schistosomiasis mansoni: skin test and cord blood reactivity to schistosomal antigens. *Trans. R. Soc. Trop. Med. Hyg.* **72:**605–609.

146. **Takemori, T., and K. Rajewsky.** 1984. Mechanism of neonatally induced idiotype suppression and its relevance for the acquisition of self-tolerance. *Immunol. Rev.* **79:**103–117.

147. **Tweardy, D. J., G. S. Osman, A. El Kholy, and J. J. Ellner.** 1987. Failure of immunosuppressive mechanisms in human *Schistosoma mansoni* infection with hepatosplenomegaly. *J. Clin. Microbiol.* **25:**768–773.

148. **UytdeHaag, F., I. Claassen, H. Bunschoten, H. Loggen, T. Ottenhoff, V. Teeuwsen, and A. Osterhaus.** 1987. Human anti-idiotypic T lymphocyte clones are activated by autologous anti-rabies virus antibodies presented in association with HLA-DQ molecules. *J. Mol. Cell. Immunol.* **3:**145–155.

149. **Velge-Roussel, F., C. Verwaerde, J. M. Grzych, and A. Capron.** 1989. Protective effects of anti-antiidiotypic IgE antibodies obtained from an IgE monoclonal antibody specific for a 26-kilodalton *Schistosoma mansoni* antigen. *J. Immunol.* **142:**2527–2532.

150. **von Lichtenberg, F.** 1987. Consequences of infections with schistosomes, p. 185–232. *In* D. Rollinson and A. J. G. Simpson (ed.), *The Biology of Schistosomes from Genes to Latrines.* Academic Press, Inc. (London), Ltd., London.

151. **Warren, K. S.** 1982. The secret of the immunopathogenesis of schistosomiasis: in vivo models. *Immunol. Rev.* **61:**189–213.

152. **Weil, G. J., R. Hussain, V. Kumaraswami, S. P. Tripathy, K. S. Phillips, and E. A. Ottesen.** 1983. Prenatal allergic sensitization to helminth antigens in offspring of parasite-infected mothers. *J. Clin. Invest.* **71:**1124–1129.

153. **Weiler, I. J., E. Weiler, R. Springer, and H. Cosenza.** 1977. Idiotype suppression by maternal influence. *Eur. J. Immunol.* **7:**591–597.

154. **Wikler, M., C. Demeur, G. Dewasme, and J. Urbain.** 1980. Immunoregulatory role of maternal idiotypes. Ontogeny of immune networks. *J. Exp. Med.* **152:**1024–1035.

Chapter 5

Anti-Idiotypic Antibodies to Bacterial Capsular Polysaccharides

M. A. J. Westerink, E. Muller, and M. A. Apicella

Despite the development of effective antibiotics and polysaccharide vaccines which appear efficacious in adults and older children (1, 14), the polysaccharide-encapsulated bacteria, including *Haemophilus influenzae, Escherichia coli,* and *Neisseria meningitidis,* remain the most common cause of serious infections in infants and young children (2, 15, 30). The immune response to polysaccharide antigens has several characteristics which distinguish it from the antibody response to protein antigens. These characteristics include thymus independence (27), failure to stimulate a memory response and to undergo affinity maturation (i.e., lack of booster response) (12), and late development in ontogeny (12, 17).

Efforts to overcome neonatal unresponsiveness to polysaccharide antigens have resulted in vaccines in which the polysaccharide capsular antigens are changed into functional T-cell-dependent antigens by coupling the polysaccharides to an immunogenic thymus-dependent carrier (6, 8, 34). An alternative strategy for the conversion of the thymus-independent polysaccharide into a thymus-dependent immunogen is the development of the surrogate image of the polysaccharide in the combining site of an anti-idiotype antibody (21). This latter technique is based on the immune regulatory network theory of Jerne (16) that defines the immune system as a web of interacting idiotopes. Antibodies carry idiotopes, which are regions in or near the antigen recognition sites. Idiotypes are capable of acting as antigens

M. A. J. Westerink • Department of Medicine, State University of New York at Buffalo School of Medicine, Buffalo, New York 14215. **E. Muller** • Department of Microbiology, State University of New York at Buffalo School of Medicine, Buffalo, New York 14215. **M. A. Apicella** • Department of Medicine and Department of Microbiology, State University of New York at Buffalo School of Medicine, Buffalo, New York 14215.

and of stimulating antibody production. The normal immune system features an interlocking network of antibodies directed at one another's idiotopes. Production of any particular antibody will be suppressed by its corresponding anti-idiotype antibody. Inherent in the network theory is the idea that at least some of the anti-idiotype antibodies are internal images of external antigens, i.e., that certain anti-idiotype antibodies express structures which resemble the structures of external antigens. For an antigen and an anti-idiotype antibody to bind to the same region of an antibody, they must have similar three-dimensional configurations even though they may be different biochemically.

BACTERIAL ANTIGENS IMPORTANT IN PATHOGENESIS

The bacterial surface provides a mosaic of structures to the immune system. Several of these structures have been isolated and used as protective vaccines. Some (*H. influenzae* type b capsular polysaccharide) have shown limited efficacy owing to the nature of the immune response they evoke (T independent) or because of the broad antigenic heterogeneity of the structure (gonococcal pili). Because of the T-independent characteristic of the immune response of the bacterial capsular polysaccharides and because they have been shown to be excellent candidates as protective vaccine for a number of gram-negative and gram-positive human pathogens, they would be logical nominal antigens for the development of anti-idiotypes.

Bacterial capsular polysaccharides have been used as vaccines since the early 1940s. These antigens are serotypic in nature and confer protection by stimulating the serum antibodies which are capable, in the presence of complement, of causing bacteriolysis and/or opsonization. The serotypic nature of these antigens requires that multivalent vaccines for a single species be developed. The current pneumococcal vaccine contains 23 different capsular polysaccharides. More than 60 other pneumococcal capsular serotypes exist but are less common causes of human infections. There are 13 different meningococcal capsular polysaccharide serogroups, and 4 (A, C, W-135, and Y) have been incorporated into a vaccine (14, 25). The meningococcal group B polysaccharide is not contained in the current vaccine because the polysaccharide is not immunogenic in humans. Serogroup B meningococci are the most frequent causes of meningococcal disease in the developed world (30).

Besides the marked heterogeneity of these capsular antigens, they have a limitation in their application based on the nature of the immune response they evoke (12, 13). They are thymus-independent antigens (9). When they are used as immunogens, an immunoglobulin M (IgM) response is evoked and switching to IgG does not occur. The response to them may be suppressed or amplified by T cells, but for the most part this response is independent of T-cell regulation (2, 28). In this regard, an anamnestic response does not occur with reimmunization with the same capsular antigen.

The T-independent response limits the usefulness of these vaccines in children younger than 2 years (12, 13, 32, 37). This is unfortunate because the majority of serious *H. influenzae* type b and meningococcal infections occur in these children. Respiratory disease in infants in developing nations is frequently secondary to infection by encapsulated pneumococci. Hence, vaccine benefits from these currently available immunogens are not available to the group at greatest risk for infection. In older children and adults the available meningococcal and the *Haemophilus* capsule generate anticapsular antibodies which are protective against the serotypes to which they have been generated.

To stimulate protective antibodies in this group of young children, it has been proposed that by conjugation of the polysaccharide antigens to proteins, vaccines could be developed which would contain a polysaccharide antigen whose immune regulation was similar to a thymus-dependent antigen (5, 6). Recently, such a vaccine has been approved for use in the United States. This vaccine is composed of the *H. influenzae* type b polysaccharide and a portion of the diphtheria toxin conjugated covalently. The vaccine requires three immunizations for optimal generation of anticapsular antibody and appeared to be effective in preventing infections among infants in studies in Finland. Other polysaccharide-protein conjugates containing capsular polysaccharides from pneumococci and meningococci are currently being developed.

DEVELOPMENT OF Ab$_2$ TO BACTERIAL ANTIGENS

Nisonoff and Lamoyi (29) and Roitt et al. (33) suggested in 1981 that internal-image antibodies might be exploited as a new type of vaccine. Monoclonal antibody technology is used to produce protective antibodies to antigenic material. These protective antibodies are subsequently purified and used to produce anti-idiotype antibodies which will be the image of the original antigen. When used as a vaccine, the anti-idiotype antibody will elicit protective immunity to the original pathogen.

Data collected over the past 10 years show that anti-idiotype antibodies directed at the hypervariable site of antimicrobial antibodies can be used to induce neutralizing antibodies in adult animals (18, 26, 31, 39) or to prime immunologically immature animals (36). Collectively, these data strongly suggest that B-cell-defined anti-idiotypes which mimic antigen may be potentially useful as antimicrobial vaccines. Anti-idiotype antibody-based vaccines may represent viable vaccine candidates in situations when it is difficult to obtain adequate amounts of purified antigen or when a conventional attenuated vaccine has a high propensity for reverting to a virulent form.

In addition, anti-idiotype antibody-based vaccines may be particularly beneficial in achieving protective immunity against encapsulated bacteria in immunologically incompetent hosts, such as neonates, who are incapable of responding to currently licensed capsular polysaccharide vaccines.

Several investigators have developed idiotope-based vaccines against bacterial pathogens and have demonstrated their usefulness in priming or inducing immunity against these pathogens.

McNamara et al. (26) showed that immunization with a monoclonal anti-idiotype antibody produced against the T15 idiotype, which is specific for phosphorylcholine, was capable of inducing high-titer, anti-phosphoryl-choline antibodies which protected mice against *Streptococcus pneumoniae* infection. The anti-idiotype antibody alone, however, was not effective in inducing protective antibodies but had to be coupled to a carrier protein, keyhole limpet hemocyanin (KLH).

Stein and Soderstrom (36) developed a monoclonal anti-idiotype antibody which mimics the capsular polysaccharide of *E. coli* K-13. Administration of the anti-idiotype antibody in conjunction with complete Freund adjuvant could prime neonatal mice and significantly enhance protection against a lethal infection with *E. coli* K-13.

We have recently developed a monoclonal antibody, 1E4, which recognizes a conserved epitope which is a target for bactericidal antibody and is present on the capsular polysaccharide of all strains of *N. meningitidis* group C tested. This monoclonal idiotype was used to produce syngeneic monoclonal anti-idiotypes (40). Not all anti-idiotypes are so-called internal images. Anti-idiotypes are classified by their physical relation to the antigen-binding site (22). Anti-idiotype antibody alpha ($Ab_2\alpha$) molecules are directed against idiotopes which are on the variable antibody domain and which are not near the antigen-binding site. Anti-idiotype antibodies which can induce steric interference with antigen binding, i.e., which are directed against an idiotope near the binding site, are called $Ab_2\gamma$. Both $Ab_2\alpha$ and $Ab_2\gamma$ are subject to genetic restriction. Anti-idiotypes which serve as surrogates of the nominal antigen are capable of competing with the nominal antigen for the binding site are called $Ab_2\beta$ (7). The binding of $Ab_2\beta$ to Ab_1 mimics the binding of antibody to antigen and is not genetically restricted. It is therefore important to correctly identify $Ab_2\beta$. In initial screening of the anti-idiotype hybridomas, a total of six hybridomas were identified which recognized the idiotype antibody (Ab_1). Only one of these anti-idiotypes, monoclonal antibody 6F9, proved to be capable of competing with (i.e., inhibiting the binding site of) the nominal antigen in inhibition enzyme-linked immunosorbent assay.

These results suggest that 6F9 recognizes an idiotypic determinant very close to, if not identical to, the paratopic regions of the combining site. Two assays were designed to test whether the specificity of 6F9 was directed at a region near the 1E4-binding site ($Ab_2\gamma$) or at the antigen-binding site ($Ab_2\beta$). For this purpose, we used polyclonal anti-meningococcal group C polysaccharide serum raised in different species in which it is unlikely that idiotypes unrelated to antigen binding would be identical. Monoclonal anti-idiotype 6F9 was capable of inhibiting the interaction of human anti-meningococcal group C polysaccharide antibodies with antigen in a solid-phase assay.

An indirect immunofluorescence technique which was previously demonstrated to be useful in identifying appropriate anti-idiotype antibodies in a hepatitis B antigen model (38) was also used. Monoclonal antibody 6F9 showed circumferential staining with rabbit anti-meningococcal group C polysaccharide typing serum and failed to stain with rabbit meningococcal group A polysaccharide serum. Furthermore, 6F9-KLH conjugate could evoke an IgG antibody response specific for the meningococcal group C polysaccharide in rabbits. These data indicate that internal-image-bearing anti-idiotype antibodies do not appear to be genetically restricted. A few other studies (11, 20) have demonstrated the induction of Ab_3 by using monoclonal anti-idiotype antibodies across species barriers. Finally, immunization of mice with 6F9 hybidoma cells resulted in the production of monoclonal Ab_3 antibodies which have the same specificity as Ab_1, thus completing the network theory of Jerne (16). These data confirm the fact that 6F9 is an $Ab_2\beta$ and functions as a surrogate antigen.

CONVERSION OF A T-INDEPENDENT TO A T-DEPENDENT ANTIGEN

Idiotype vaccines are proteins and therefore may act as T-dependent antigens. Only two groups (10, 35) have shown that an anti-idiotype antibody vaccine is able to induce a B- and T-cell-mediated response. An anti-idiotype antibody which mimicked a T-dependent antigen was used in both systems. McNamara et al. (26) and Stein and Soderstrom (36) have previously shown the usefulness of anti-idiotype antibody vaccines by inducing and priming for protective antibodies against encapsulated organisms. However, these investigators did not report the characteristics of the immune response. In addition, it was necessary to either couple the anti-idiotype antibody to a protein carrier or use an adjuvant to induce these protective antibodies. The use of adjuvants and/or carrier molecules seemed to be a prerequisite in many instances for optimal antibody production by anti-idiotypes (11, 19).

Several investigators (E. Muller, personal communication; M. McNamara Ward, personal communication) have obtained experimental data which suggest that certain anti-idiotype antibodies which mimic T-independent antigens will elicit a T-independent and not a T-dependent antibody response. The exact nature of the immune response elicited by anti-idiotype antibodies which mimic T-independent antigens such as capsular polysaccharides remains to be elucidated.

The data regarding the immune response to T-independent antigens are based on the studies with type III pneumococcal polysaccharide (3). In similar studies, the T-independent nature of the immune response has been demonstrated for *H. influenzae* type b polysaccharide (23), meningococcal group A and C polysaccharides (4, 28) and dextran (24). Using a similar murine model, we have investigated the nature of the immune response evoked by meningococcal group C polysaccharide and by an anti-idiotypic antibody which mimics the group C capsular polysaccharide.

A dose-response curve was generated both for the nominal antigen, meningococcal group C polysaccharide, and for the surrogate antigen, the anti-idiotype antigen 6F9, in BALB/c mice. The optimal dose of meningococcal group C polysaccharide was 5 µg per mouse, and the optimal dose of 6F9 was 100 µg per mouse. To study the kinetics of the antibody response, we immunized BALB/c mice with an optimal dose of either meningococcal group C polysaccharide or 6F9. The anti-meningococcal group C polysaccharide IgM and IgG antibody titers were determined on a weekly basis. Mice immunized with meningococcal group C polysaccharide generated a group C-specific IgM response, which was maximal by day 7 and remained constant over the 4-week period of study (Fig. 1A). Mice immunized with 6F9 also generated a group C-specific IgM response, which was maximal at day 7 but then decreased significantly. Mice immunized with meningococcal group C polysaccharide failed to generate an anti-meningococcal group C polysaccharide IgG response (Fig. 1B). In contrast, mice immunized with 6F9 showed a significant IgG response, which was 100-fold greater than the background. This IgG response was first detectable at day 14 and peaked at day 28.

Studies with concanavalin A and antilymphocyte serum showed that prior treatment with concanavalin A suppresses the immune response to meningococcal group C polysaccharide, whereas antilymphocyte serum enhances this immune response. This is characteristic of a T-independent antigen. Conversely, in animals immunized with 6F9, treatment with concanavalin A results in an enhancement of the immune response, suggesting stimulation of T-amplifier cells, while antilymphocyte serum suppresses the immune response.

These results strongly suggest that 6F9, the surrogate image of a T-inde-

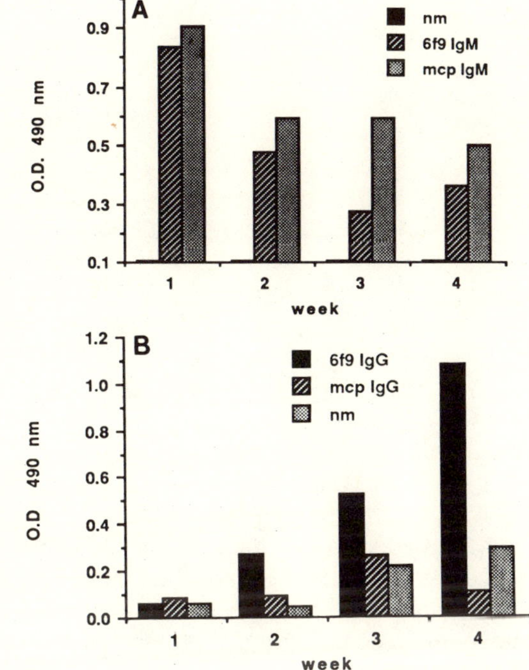

Figure 1. Enzyme-linked immunosorbent assay of mouse sera for anti-meningococcal group C polysaccharide (mcp) antibodies. BALB/c mice (five per group) were immunized intraperitoneally with 5 µg of meningococcal group C polysaccharide or 100 µg of 6F9. The anti-meningococcal group C polysaccharide IgM (A) and the anti-meningococcal group C polysaccharide IgG (B) responses were measured on a weekly basis. Microdilution wells were coated with 10 µg of meningococcal group C polysaccharide per ml.

pendent antigen, acts as a T-dependent antigen. Athymic nu/nu mice, incapable of responding to T-dependent antigens, were used to confirm these findings. The results of these studies showed that nu/nu mice did not respond to 6F9, whereas the $nu/+$ controls showed a normal anti-meningococcal group C polysaccharide antibody response. The nu/nu mice immunized with meningococcal group C polysaccharide were capable of responding to this T-independent antigen, as were the $nu/+$ controls.

As one would expect from a T-dependent antigen, a significant secondary response or booster effect was seen in mice that received a second immunization with 6F9. The anti-meningococcal group C polysaccharide IgG antibody response of the mice that received a booster dose of 6F9 was fivefold higher than the response of the mice that received a single dose of 6F9. This booster response was not detected in the mice immunized with meningococcal group C polysaccharide. Furthermore, mice immunized with 6F9 and subsequently boosted with live group C meningococci showed a similar booster response (Fig. 2). The mice immunized with 6F9 and subsequently boosted with live meningococci showed a three- to fivefold increase in the specific anti-meningococcal group C polysaccharide IgG response. Mice immunized with meningococcal group C polysaccharide

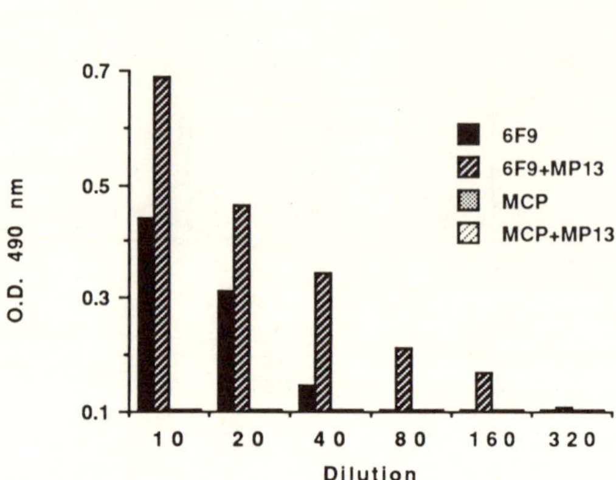

Figure 2. Enzyme-linked immunosorbent assay of mouse sera for anti-meningococcal group C polysaccharide IgG antibody response. BALB/c mice (five per group) were immunized with 5 µg of meningococcal group C polysaccharide (MCP) or 100 µg of 6F9 on day 0 and received a booster injection with 10^3 CFU of live group C meningococci intraperitoneally on day 21. The control group of mice consisted of mice immunized with 5 µg of meningococcal group C polysaccharide or 100 µg of 6F9 on day 0. The anti-meningococcal group C polysaccharide IgG antibody response was measured on day 26.

and boosted with live meningococci did not develop a secondary anti-meningococcal group C polysaccharide IgG response. These data suggest that the subset of T memory cells stimulated by 6F9 are the T cells involved in the meningococcal group C polysaccharide response when confronted with live group C meningococci.

Finally, data obtained from neonatal mice (8 to 10 days old) that received a primary and secondary injection of either meningococcal group C polysaccharide or 6F9 indicate that these neonatal mice can be primed effectively with the anti-idiotype antibody (Fig. 3). Mice immunized and boosted with meningococcal group C polysaccharide did not have a detectable anti-meningococcal group C polysaccharide IgG response, whereas mice immunized and boosted with 6F9 showed a significant anti-meningococcal group C polysaccharide IgG response which was comparable to the IgG response seen after 4 weeks in adult mice immunized with 6F9.

We did not find it necessary to use carrier proteins or adjuvants in these experiments. Other investigators (11, 19) have found the use of carrier molecules and/or adjuvants necessary. In some of these cases a suboptimal dose of anti-idiotype antibody may also have been used. In addition, we speculate that a number of anti-idiotype antibody molecules may require a carrier molecule and/or adjuvant for correct or effective antigen presentation and processing. Finally, the location of the surrogate image on the immunoglobulin molecule (variable heavy versus variable light chain, or a combination of both) may play a crucial role in this phenomenon. Recent advances in immunoglobulin and mRNA sequencing techniques will enable us to clarify these issues.

Figure 3. Enzyme-linked immunosorbent assay of neonatal-mouse sera for anti-meningococcal group C polysaccharide IgG antibody response. Neonatal BALB/c mice (five per group) 8 to 10 days of age were immunized with 1 µg of meningococcal group C polysaccharide (MCP) or 50 µg of 6F9 intraperitoneally. On day 21 these mice received either no booster injection (MCP-1x and 6F9-1x) or 5 µg of meningococcal group C polysaccharide (MCP-2x) or 100 µg of 6F9 (6F9-2x). The anti-meningococcal group C polysaccharide IgG response was measured on day 26. Microdilution wells were coated with 10 µg of meningococcal group C polysaccharide per ml.

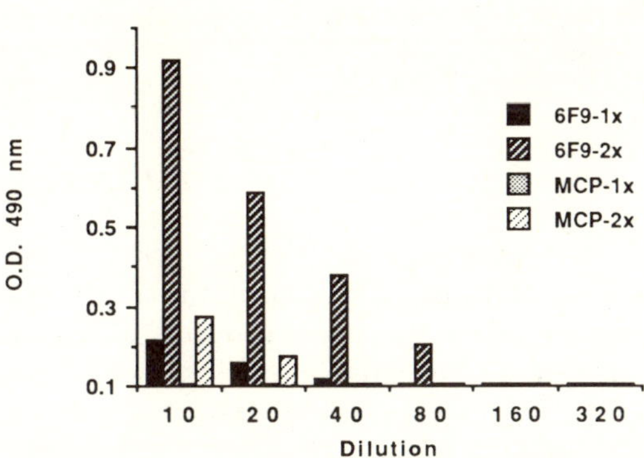

ANTI-IDIOTYPE IMAGES AS T-INDEPENDENT IMMUNOGENS

A series of anti-idiotype antibodies were developed by using the meningococcal group A capsular polysaccharide as the nominal antigen. Several of these antibodies have been studied, and these investigations indicate that at least one of them elicits an immune response which is T independent and identical to that of the nominal polysaccharide antigen. This antibody, AC8, was generated after immunization with the monoclonal antibody 10G5, which has specificity for the meningococcal group A polysaccharide. This anti-idiotype antibody recognized a determinant(s) on Fab fragments of 10G5 and inhibited antipolysaccharide IgM and IgG from BALB/c, NZB, and outbred Swiss Webster mice. A single dose of the anti-idiotype antibody AC8 coupled to KLH elicited anti-polysaccharide IgM in syngeneic and allogeneic mice. Low levels of anti-polysaccharide IgG were also elicited by multiple doses of AC8-KLH in adjuvant, but the levels of the anti-polysaccharide IgG could not be augmented by modifications of the immunization protocol or by administration of additional doses of the conjugate in adjuvant.

Although rat and rabbit anti-meningococcal group A polysaccharide antibodies specifically bound AC8 and were inhibited by it, neither species responded with anti-polysaccharide antibodies to multiple doses of AC8-KLH in adjuvant. However, both species failed to respond or responded poorly to single or multiple doses of purified polysaccharide given with or without adjuvant. Both species produced anti-polysaccharide IgM and IgG following a single dose of live group A meningococci. Thus, whereas anti-idiotype antibody mimicry of the meningococcal serogroup A polysaccharide, a T-independent antigen, is possible, it appears that the same restrictions which apply to humoral responsiveness to the polysaccharides may also apply to the response against AC8, an anti-idiotype antibody which contains a surrogate image of the meningococcal group A polysaccharide.

CONCLUSIONS

In summary, we have developed an anti-idiotypic antibody, 6F9, whose combining site contains a surrogate image of a T-independent antigen, meningococcal group C polysaccharide; acts as a T-dependent antigen; and is capable of eliciting an anti-meningococcal group C polysaccharide IgG antibody response. In addition, 6F9 evokes a booster response classically seen with T-dependent antigens and is capable of priming neonatal animals that do not respond to T-independent antigens. A carrier molecule or adjuvant was not necessary to elicit an antibody response. This is typically the kind of response expected from anti-idiotype-based vaccines, which are proteins and therefore should act as T-dependent agents. However, we also have presented data from our studies involving an anti-idiotypic antibody, AC8, whose combining site contains a surrogate image of another T-independent antigen, the meningococcal A polysaccharide, which acts as a T-dependent antigen in various animal models. Despite the use of both a carrier molecule and an adjuvant, AC8 failed to stimulate an IgG response or a booster effect.

From these studies, we conclude that not all anti-idiotypic antibodies act as T-dependent antigens, but that certain anti-idiotypic antibodies may remain T-independent antigens despite their chemical nature. It will therefore be necessary to screen each anti-idiotype antibody carefully and to define the nature of the immune response elicited by each anti-idiotype antibody considered to be a vaccine candidate. The fact that not all anti-idiotype antibodies are T-dependent antigens should not discourage anti-idiotype antibody vaccine development, since the amino acid sequence of these immunoglobulins can be obtained using the mRNA of the hybridoma cells and converted into synthetic peptides which may be coupled to various potent carrier molecules and become T-dependent antigens.

Finally, these T-independent and T-dependent anti-idiotypes may allow us to study and increase our understanding of what determines the T-independent or T-dependent nature of antigens.

Acknowledgments. This work was supported by Public Health Service grants AI26279 and AI18384 from the National Institutes of Health and by the Ralph Hochstetter Medical Research Fund in honor of Henry C. and Bertha H. Buswell.

REFERENCES

1. **Artenstein, M. S., R. Gold, J. G. Zimmerly, F. A. Wyle, W. C. Branche, Jr., and C. Harkins.** 1970. Prevention of meningococcal disease by group C polysaccharide vaccine. *N. Engl. J. Med.* **282:**417–420.

2. **Austrian, R.** 1981. Some observations on the pneumococcus and on the current status of pneumococcal disease and its prevention. *Rev. Infect. Dis.* **3:**S1–S17.

3. **Baker, P. J., D. F. Amsbaugh, P. W. Stashak, G. Caldes, and B. Prescott.** 1981. Regulation of the antibody response to pneumococcal polysaccharide by thymus-derived cells. *Rev. Infect. Dis.* **3:**332–341.

4. **Beuvery, E. C., A. B. Leussink, R. W. van Delft, R. H. Tiesjema, and J. Nagel.** 1982. Immunoglobulin M and G responses and persistence of these antibodies in adults after vaccination with a combined meningococcal group A and group C polysaccharide vaccine. *Infect. Immun.* **37:**579–585.

5. **Beuvery, E. C., R. W. van Delft, F. Medema, V. Kanhai, and J. Nagel.** 1983. Immunological evaluation of meningococcal group C polysaccharide-tetanus toxoid conjugate in mice. *Infect. Immun.* **41:**609–617.

6. **Beuvery, E. C., F. Van Rossum, and J. Nagel.** 1982. Comparison of the induction of immunoglobulin M and G antibodies in mice with purified pneumococcal type 3a and meningococcal group C polysaccharides and their protein conjugates. *Infect. Immun.* **37:**15–22.

7. **Bona, C. A., and H. Kohler.** 1984. Anti-idiotypic antibodies and internal images, p. 141–149. *In* C. J. Venter, C. M. Fraser, and J. Lindstrom (ed.), *Receptor Biochemistry and Methodology,* Vol. 4. Alan R. Liss, Inc., New York.

8. **Braley-Muller, H.** 1980. Antigen requirements for priming of IgG producing B memory cells specific for type III pneumococcal polysaccharide. *Immunology* **40:**521–527.

9. **Brandt, B., and M. S. Artenstein.** 1975. Duration of antibody responses after vaccination with group C Neisseria meningitidis polysaccharide. *J. Infect. Dis.* **131:**569–574.

10. **Ertl, H. C. J., and R. W. Finberg.** 1984. Sendai virus-specific T cell clones: induction of cytolytic T cells by an anti-idiotypic antibody directed against a T helper cell clone. *Proc. Natl. Acad. Sci. USA* **81:**2850–2854.

11. **Gaulton, G. N., A. H. Sharpe, D. W. Chang, B. N. Fields, and M. I. Greene.** 1986. Syngeneic monoclonal internal image anti-idiotypes as prophylactic vaccines. *J. Immunol.* **137:**2930–2936.

12. **Gold, R., and M. L. Lepow.** 1975. Clinical evaluation of group A and C meningococcal polysaccharide vaccines in infants. *J. Clin. Invest.* **56:**1536–1547.

13. **Gotschlich, E. C., M. Rey, R. Triau, and K. S. Sparks.** 1972. Quantitative determination of the human immune response to immunization with meningococcal vaccines. *J. Clin. Invest.* **51:**89–96.

14. **Griffiss, J. M., B. L. Brandt, and D. O. Broud.** 1982. Human immune response to various doses of group Y and W135 meningococcal polysaccharide vaccines. *Infect. Immun.* **37:**205–208.

15. **Hill, J. C.** 1983. Summary of a workshop on *Haemophilus influenzae* type B vaccines. From the National Institute of Allergy and Infectious Diseases. *J. Infect. Dis.* **148:**167–175.

16. **Jerne, N. K.** 1974. Toward a network theory of the immune system. *Ann. Immunol.* (Paris) **125C:**373–389.

17. **Kayhty, H., V. Karanko, H. Peltola, and P. H. Makela.** 1984. Serum antibodies after immunization with *H. influenzae* type B capsular polysaccharide and responses to reimmunization: no evidence of immunologic tolerance or memory. *Pediatrics* **74**:857–865.

18. **Kennedy, R. C., and K. Adler-Storthe.** 1983. Immune response to hepatitis B surface antigen: enhancement by prior injection of antibodies to the idiotype. *Science* **221**:853–854.

19. **Kennedy, R. C., and G. R. Dreesman.** 1984. Enhancement of the immune response to hepatitis B surface antigen. *J. Exp. Med.* **159**:655–665.

20. **Kennedy, R. C., J. W. Eichberg, R. E. Lanford, and G. R. Dreesman.** 1986. Anti-idiotypic antibody vaccine for type B viral hepatitis in chimpanzees. *Science* **232**:220–221.

21. **Kieber-Emmons, T., R. E. Ward, S. Raychaudhuri, R. Rein, and H. Kohler.** 1986. Rational design and application of idiotope vaccines. *Int. Rev. Immunol.* **1**:1–26.

22. **Kohler, H., S. Muller, and C. Bona.** 1985. Internal antigen and immune network. *Proc. Soc. Exp. Biol. Med.* **178**:189–195.

23. **Lee, C. J., F. G. Malik, and J. B. Robbins.** 1978. The regulation of the immune response of mice to *Haemophilus influenzae* type b capsular polysaccharide. *Immunology* **34**:149–156.

24. **Leon, M. A., J. C. Chen, and T. H. Kuo.** 1978. Regulatory events in the immune response of mice to dextran, p. 105–120. *In* J. A. Rudbach and P. J. Baker (ed.), *Immunology of Bacterial Polysaccharides.* Elsevier/North Holland Publishing Co., New York.

25. **Lepow, M. L., J. Beeler, M. Randolph, J. S. Samuelson, and W. A. Hankins.** 1986. Reactogenicity and immunogenicity of a quadrivalent combined meningococcal vaccine in children. *J. Infect. Dis.* **154**:1033–1036.

26. **McNamara, M., R. E. Ward, and H. Kohler.** 1984. Monoclonal idiotype vaccine against *Streptococcus pneumoniae* infection. *Science* **226**:1325–1326.

27. **Mosier, D. E., N. M. Zalvidas, E. Goldings, J. Mond, I. Scher, and W. E. Paul.** 1977. Formation of antibody in the newborn mouse: study of T-cell independent antibody response. *J. Infect. Dis.* **136**:S14–S19.

28. **Muller, E., and M. A. Apicella.** 1988. T-cell modulation of the murine antibody response to *Neisseria menigitidis* group A capsular polysaccharide. *Infect. Immun.* **56**:259–266.

29. **Nisonoff, A., and E. Lamoyi.** 1981. Hypothesis implications of the presence of an internal image of the antigen in anti-idiotypic antibodies: possible application to vaccine production. *Clin. Immunol. Immunopathol.* **21**:397–406.

30. **Peltola, H.** 1983. Meningococcal disease: still with us. *Rev. Infect. Dis.* **5**:71–91.

31. **Reagan, K. J., W. H. Wunner, I. J. Wiktor, and H. Koprowski.** 1983. Anti-idiotypic antibodies induce neutralizing antibodies to rabies virus glycoprotein. *J. Virol.* **48**:660–666.

32. **Reingold, A. L., A. W. Hightower, G. A. Bolan, E. E. Jones, H. Tiendrebeogo, A. Yada, C. Phillips, C. Adamsbaum, G. W. Ajello, and C. V. Broome.** 1985. Age specific differences in duration of clinical protection after vaccination with meningococcal polysaccharide vaccine. *Lancet* **ii**:114–118.

33. **Roitt, I. M., A. Cooke, D. K. Male, F. C. Hay, G. Guarnotta, P. M. Lydyard, L. P. de Carvalha, Y. M. Thanavala, and J. Ivanyi.** 1981. Idiotypic networks and their possible exploitation for manipulation of the immune response. *Lancet* **i**:1041–1045.

34. **Schneerson, R., O. Barrera, A. Sutton, and J. B. Robbins.** 1980. Preparation, characterization and immunogenicity of *Haemophilus influenzae* type b polysaccharide-protein conjugates. *J. Exp. Med.* **152**:361–376.

35. **Sharpe, A. H., G. N. Gaulton, K. K. McDade, B. N. Fields, and M. I. Greene.** 1984. Syngeneic monoclonal anti-idiotypic antibody can induce cellular immunity to reovirus. *J. Exp. Med.* **160**:1195–1205.

36. **Stein, K. E., and T. Soderstrom.** 1984. Neonatal administration of idiotype or anti-idiotype primes for protection against *E. coli* K13 infection in mice. *J. Exp. Med.* **160**:1001–1011.

37. **Taunay, A. de E., P. A. Galvao, J. S. de Morais, E. C. Gotschlich, and R. A. Feldman.** 1974. Disease prevention by meningococcal serogroup C polysaccharide vaccine in pre-school: results after eleven months in San Paulo, Brazil. *Pediatr. Res.* **8**:429. (Abstract.)
38. **Thanavala, Y. M., and I. M. Roitt.** 1986. Monoclonal anti-idiotypic antibodies as surrogates for hepatitis B surface antigen. *Int. Rev. Immunol.* **1**:27–39.
39. **Uytedhaag, F., and A. Osterhaus.** 1985. Induction of neutralizing antibody in mice against poliovirus type II with monoclonal anti-idiotypic antibody. *J. Immunol.* **134**:1225–1229.
40. **Westerink, M. A. J., A. A. Campagnari, M. A. Wirth, and M. A. Apicella.** 1988. Development and characterization of an anti-idiotype antibody to the capsular polysaccharide of *Neisseria meningitidis* serogroup C. *Infect. Immun.* **56**:1120–1127.

Chapter 6

Idiotypic Markers of the Immune Response to Mycobacterial Antigens

J. Ivanyi

The immunological study of mycobacterial diseases is currently being applied to several fundamental aspects of protective and pathogenic host responses to infection (17). It is generally known that the great majority of individuals infected with *Mycobacterium tuberculosis* or *M. leprae* do not develop disease because they are protected by innate and acquired resistance. Protective immunity has been attributed to CD4 cell-secreted lymphokines which activate macrophages for the bacteriocidal activity. However, CD4 T-cell assays, such as skin delayed hypersensitivity (DTH) or in vitro proliferation, must be interpreted in the light of compelling evidence against the role of DTH reactivity in protective immunity to mycobacterial infection (29). In addition to the CD4 cell-mediated immunity, it was recently proposed that CD8 cells may lyse infected target cells which are not equipped with bacteriocidal capacity; subsequently, the released bacteria would be taken up and killed by macrophages with unimpaired bactericidal function (25). Nevertheless, none of the parameters which would measure T-cell, lymphokine, or macrophage function in vitro represent an adequate immunological assay for the host protective immunity against mycobacterial pathogens. This lack of a relevant test also represents a major limitation for the exact assessment of prophylactic immunization of humans with the *M. bovis* BCG vaccine.

The molecular nature of antigens which can stimulate protective immunity is of obvious interest. Although vaccination of mice with live *M. bovis* BCG or with killed *M. leprae* can induce T-cell-mediated protective immunity toward subsequent systemic or local challenge, the degree of

J. Ivanyi • Medical Research Council Tuberculosis and Related Infections Unit, Royal Postgraduate Medical School, Hammersmith Hospital, London W12 OHS, United Kingdom.

protection imparted by vaccination is incomplete, and human trials with BCG have produced results with unsatisfactory variability between geographic regions. Considering that patients with tuberculosis or tuberculoid leprosy have abundant but ineffective T-cell immunity, the vaccination strategy should not aim merely at increasing the magnitude of the host response. Instead, it may be desirable to stimulate selectively the immunity to certain antigens or the response of a particular subset of T cells.

Major advances in the molecular study of mycobacterial antigens in recent years were initiated with monoclonal antibodies (MAb) produced since 1981 in several laboratories (22). Genomic libraries of mycobacterial DNA have enabled the sequencing and expression of an increasing number of protein constituents and the topographic study of epitopes (32). The immunogenicity of the corresponding synthetic peptides could be increased by the construction of dimeric peptides and conferred to previously nonimmunogenic peptides (10). MAbs identifying species-specific epitopes for either *M. tuberculosis* or *M. leprae* led to the possibility of developing a serodiagnostic test. A competition assay showed some of these epitopes to be immunodominant, i.e., giving a high proportion of positive sera from patients with multibacillary tuberculosis and lepromatous leprosy (18). It is of interest that when comparing several specificities, the pattern of immunodominance varies with the form of disease or infection (5, 18, 19, 24).

Immune reactions decide not only between protection or disease but also the type of clinical manifestations in leprosy. Thus, classification of leprosy has been based essentially on the balance between immune reactions, namely, the paucibacillary tuberculoid leprosy characterized by strong T-cell reactions and the multibacillary lepromatous leprosy characterized by pronounced antibody formation. The immunological defects may be due to (i) a change in the specificity of the immune repertoire, (ii) a defect in the composition or secretion of lymphokines, or (iii) a defective response of macrophages to lymphokines, which could occur at the level of antigen processing or, alternatively, as a result of making the macrophages selectively refractory for the killing of mycobacteria. T-cell anergy could be the cause of the unrestrained growth of *M. leprae,* but the alternative view that it is merely a temporal consequence of excessive bacterial expansion cannot be excluded. It was reported that CD8 cells with specificity for the phenolic glycolipid I constituent of *M. leprae* could suppress the response of CD4 cells to the polyclonal mitogen concanavalin A (3). Specific proteins from *M. leprae* were also implicated in a regulatory capacity by the demonstration that they suppressed the skin DTH reaction to tuberculin (50). However, this effect was demonstrable in patients with both lepromatous and tuberculoid leprosy and therefore agreed with the view that the reported suppressor cells could not be responsible for the immunopathogenesis of lepromatous leprosy (43). Attempts had been made to restore the immune capacity of lepromatous T cells in vitro by using exogenous interleukin-2 in antigen-stimulated lymphocyte cultures. The success rate varied between laboratories, and on the whole it appears that significant reversal can be achieved only in a minority of cases (36). Differences may exist between cells in the peripheral blood and those at the site of pathological lesions (47), and T cells trapped near to infected or antigen-containing cells become organized in granulomas in which the CD4/CD8 ratio and bacillary count varies with the form of leprosy (34).

A recent study revealed the association of class II human leukocyte antigen alleles with the incidence of pulmonary tuberculosis and with the antibody levels to a 38-kilodalton (kDa) protein antigen of *M. tuberculosis* (6). In this study antibody titers to only one of five tested antigens were associated with the DR2 allele. The study also established the association of disease with DR2 and DQw1 (36 and 39% attributable risk, respectively) and of resistance to tuberculosis with DQw3 (57% preventative fraction). The association of DR2 with specific antibody levels as well as with disease raises the possibility of a pathogenetic role of the 38-kDa antigen. It seems conceivable that the high-responder antibody phenotype is signaling a decline in the macrophage-activating, i.e., protective, immune function by T cells which had become engaged as helpers for the B-cell response. If the assumption about the reciprocal relationship between humoral and protective immunity is correct, the antibody specificity represents a clue in the search for protective antigens.

The idiotypic network of the host antimycobacterial responses is of interest for the following reasons. (i) Anti-idiotype antibodies, following injection in experimental animals, have been shown to be stimulatory for various forms (helper, cytotoxic, DTH) of T-cell immunity (12, 35, 52, 53, 55). Thus, anti-idiotypic reagents may be valuable in the analysis of the distinct immunological defects in mycobacterial diseases. (ii) When considering subunit vaccines (11) anti-idiotypic antibodies may give the image of conformational epitopes (33), based on nonlinear sequences; these are difficult to produce by using recombinant DNA or synthetic peptide technologies. (iii) Immune responses to distinct epitopes of mycobacteria may be protective or pathogenic in various individuals on the basis of genetic restrictions. It may be possible to overcome these restrictions with anti-idiotypic antibodies (48).

PRIVATE IDIOTYPES

The analysis of idiotypic determinants was undertaken with collections of MAbs which had been raised against antigens of *M. tuberculosis* and *M. leprae* (8, 20, 23). These MAbs (Table 1) are directed to several species-spe-

Table 1. Review of Anti-Idiotype Antibodies[a]

Code of rabbit anti-idiotype	MAb (immuno-globulin)	Antigen	Specificity of MAb			Idiotype expression	Anti-idiotype titer
			M. leprae	*M. tuberculosis*	Others		
Rb06	ML06 (IgG1)	12 kDa	+	−	−	H chain	8,000
Rb10	ML10 (IgG2a)[b]					Whole Ig[c]	5,000
Rb04	ML04 (IgG1)[d]	35 kDa	+	−	−	Whole Ig	10,000
Rb38	ML38 (IgG1)					Whole Ig	3,500
Rb30	ML30 (IgG1)	65 kDa	+	+	Broad	Whole Ig	20,000
Rb34	ML34 (IgM)	LAM[e]	+	+	Broad	Whole Ig	3,500
Rb02	ML02 (IgG3)					Whole Ig	600
Rb23	TB23 (IgG1)	19 kDa	−	+	Limited	Whole Ig	700
Rb78	TB78 (IgG1)	65 kDa	−	+	Limited	L chain	2,700
Rb68	TB68 (IgG1)	14 kDa	−	+	*M. bovis*	Whole Ig	100
Rb71	TB71 (IgG2b)	38 kDa	−	+	−	Whole Ig	700
Rb72	TB72 (IgG1)					Whole Ig	500

[a] From reference 40.
[b] Overlapping idiotype on ML05 (IgG2a) and ML12 (IgG2b).
[c] Ig, Immunoglobulin.
[d] Overlapping idiotype on ML03 (IgG1) and ML11 (IgG1).
[e] LAM, Lipoarabinomannan.

cific and cross-reactive antigens which had been analyzed from various aspects of the immune response to mycobacterial infection (22). Anti-idiotypic antibodies were produced by a conventional procedure in rabbits: sodium sulfate-precipated globulins from ascitic fluids were injected into New Zealand White rabbits in incomplete Freund adjuvant (IFA), and antisera were extensively absorbed by normal mouse immunoglobulin-coupled CNBr-Sepharose 4B column chromatography. The idiotypic specificity of absorbed sera was established on the basis of binding immunoassays with homologous MAb-immunoglobulin and the lack of binding with heterologous MAb-immunoglobulin- or normal mouse immunoglobulin-coated polyvinyl microdilution plates. Competition immunoassays were done with anti-idiotype-coated plates. The idiotypes of two groups of four MAbs, each of overlapping paratope specificity toward the 12- and 35-kDa antigens, specific for *M. leprae* have been analyzed. In each of these groups, two distinct idiotypes, carried by MAbs, of apparently overlapping paratope specificity have been identified (41). Immunoblot analysis of separated chains showed that most of the idiotypes were detected only on intact immunoglobulin molecules, with the exception of one H-chain (Rb06) and one L-chain (Rb78) specificity. None of the 12 idiotypes listed in Table 1 showed any significant cross-reactivity, suggesting that they can be classified as private specificities. The position of idiotypes in relation to the antibody-combining site (paratope) was determined by competition with anti-idiotypes with the binding of labeled MAb-immunoglobulin to antigen-coated plates. All 12 anti-idiotypes competed with the binding of MAbs to antigen-coated wells, indicating that the corresponding idiotypes are localized near the antibody-combining site. However, MAbs showed 10- to 100-fold-higher potency in the idiotype than in the paratope competition assay, indicating either that idiotype–anti-idiotype binding is of higher affinity than paratope-antigen binding or that the idiotypes are topographically placed in the periphery of the antibody-combining site. Furthermore, considerable differences in the affinity of idiotype–anti-idiotype binding between the examined specificities have been identified by dissociation assays (41).

The representation within the polyclonal antibody response of antigen-immunized mice was examined only for the Rb04 idiotype. Quantitative serological results indicated a paucity of this idiotype within the polyclonal repertoire, since the idiotype was detected at a ca. 100-fold-lower serum dilution than the corresponding paratope specificity (41). Despite limitations in the sensitivity of detection, the analysis of congenic mouse strains indicated an association of the Rb04 idiotype-positive response with the *Igh* locus (low response of Igh-1 [b]) and no association with the *H-2* locus. However, a survey of sera from patients with lepromatous leprosy which contained high levels of ML04 paratope-positive antibodies showed an absence of the corresponding Rb04 idiotype (40). A similar analysis of sera from patients with pulmonary tuberculosis indicated an absence of the RB72 and RB71 idiotypes on the anti-38-kDa antibodies. Although a more comprehensive analysis may be required, it appears from the present data that the antibody repertoire following immunization or infection with mycobacterial pathogens is generated by extensive somatic diversification and is therefore reflected by a very heterogeneous collection of private idiotypic specificities. The existence of common idiotypic marker on these antibodies cannot be categorically excluded, but its occurrence as found in the case of autoantibodies (see Public Idiotypes) has so far not been detected.

ANTI-IDIOTYPIC STIMULATION OF MOUSE B CELLS

Injection of anti-idiotypic antibodies to animals has often induced an idiotypic response (Ab$_3$) of homologous idiotypic specificity, but demonstration of antigen-binding activity corresponding to the internal-image specificity has not been a regularly predictable outcome. Although antibodies of the correct specificity and even protective potency have been reported in respect to various bacterial, viral, and parasitic infections (11, 35), the lack of antigen binding by the Id^+ Ab_3 antibodies in certain systems has been well documented (4, 38, 54).

The biological effects of two anti-idiotypes relating to the 35- and 12-kDa *M. leprae*-specific antigens (Rb04 and Rb10) have been investigated (41). BALB/c mice were injected first with anti-idiotype antibody in IFA and then challenged either with the homologous anti-idiotype or with antigen (Table 2). An Id^+ serum immunoglobulin response, strictly idiotype specific, was detected 4 to 6 weeks after injection of either Rb04 or Rb10. However, the paratope (antigen-binding) activity was not demonstrable following anti-idiotype injections alone and was not amplified in mice injected with anti-idiotype and antigen in sequence. In fact, cross-absorption analysis unequivocally demonstrated that the idiotype- and paratope-positive (P^+) sera from doubly immunized mice contained two nonoverlapping and mutually exclusive populations of Id^+ (RB04) and P^+ immunoglobulin molecules, respectively. The dissociation between these two functions was also apparent for the ML10/Rb10 specificities. This epitope (unlike ML04) is nonimmunogenic following antigenic stimulation of mice, and it is also silent in patients with lepromatous leprosy whose sera contain high ML04-specific antibody levels. As in the case of Rb04, injection of Rb10 induced a pronounced Id^+ Ab_3 serum response, but completely failed to induce or prime for a paratope-positive antibody response.

Table 2. Stimulation of Paratope-Negative Idiotype by Anti-Idiotype[a]

Immunization of mice		Expression of:			
		Paratope		Idiotope	
Priming (+IFA)	Challenge (soluble)	ML04	ML10	Rb04	Rb10
Rb04	Rb04	−	−	+++	−
	MLSE[b]	+++[c]	−	+++[d]	−
Rb10	Rb10	−	−	−	++
	MLSE	++	−	−	+++
NRS[e]	Rb04	−	−	−	−
	MLSE	++	−	−	−

[a] From reference 41.
[b] MLSE, *M. leprae* soluble extract.
[c] Antibody activity absorbed by *M. leprae* soluble extract but not by Rb04 column.
[d] Antibody activity absorbed by Rb04 but not by *M. leprae* soluble extract column.
[e] NRS, Normal rabbit serum.

The Id$^+$ responses in the experiments reviewed above were dependent on the inoculation of the rabbit anti-idiotypes in IFA, confirming the adjuvant dependency of the Ab$_3$ stimulation reported previously (15, 57). Moreover, it was found that injection of soluble anti-idiotype produced a cyclophosphamide-sensitive suppression of the Ab$_3$ response to a subsequent injection of Rb04 anti-idiotype in IFA (Table 3). This finding suggests that the Ab$_3$ response, which is known to be T-cell dependent (2), can be modulated in a similar way as murine immune responses to protein antigens in general (56). This mechanism may also have been responsible for earlier observations of idiotype suppression by injection of anti-idiotypic antibodies in soluble form (16, 26).

ANTI-IDIOTYPIC STIMULATION OF MOUSE T CELLS

It has been demonstrated previously that anti-idiotypic antibodies raised against immunoglobulin idiotypes can evoke T-cell-mediated reactions such as DTH, cytotoxicity, or help for antibody formation (12, 52, 53, 55). Since T cells play a mandatory role in both protective and pathogenic interactions in the infected host, the potential of anti-idiotype reagents to display an internal image of T-cell-stimulatory epitopes of mycobacterial antigens has been of considerable interest. Of the several anti-idiotype reagents listed in Table 1, only antibody Rb71 has significant biological activities. By using antigen-sensitized mice, affinity-purified Rb71 globulin stimulated proliferative in vitro responses of T cells of the Lyt.1 subset (Table 4) (42). Moreover, anti-idiotype elicited significant DTH footpad reactions in BALB/c mice presensitized with the *M. tuberculosis* soluble extract. The immunogencity of Rb71 in vivo was demonstrable following injection in

Table 3. Cyclosphosphamide-Sensitive Inhibition of the Id$^+$ P$^-$ Response to Anti-Idiotype/IFA by Fluid Anti-Idiotype[a]

Group	Cyclophosphamide (day −2)	Rb04 injection at:		Relative response to:	
		0 + 4 days	21 days	Rb04 idiotype	NRbG[b]
1	−	−	IFA	100	100
2	−	Fluid	IFA	0	96
3	+	Fluid	IFA	111	123
4	+	−	IFA	78	NT[c]
5	−	−	−	0	0

[a] Revised from reference 41.
[b] NRbG, Normal rabbit globulin.
[c] NT, Not tested.

Table 4. T-Cell Responses of BALB/c Mice to Anti-Idiotype Rb71[a]

Priming	Elicitation	[^{3}H]thymidine uptake (10^{-3} cpm)	Footpad DTH (10^{-2} mm)
MTSE[b]	MTSE	51.6	77
	Anti-idiotype Rb71	21.5	51
	Normal immunoglobulin	0.9	17
Anti-idiotype Rb71	MTSE	35.5	84
	38-kDa antigen	18.3	NT[c]
	Saline	5.7	<10

[a] Revised from reference 42.
[b] MTSE, *M. tuberculosis* soluble extract.
[c] NT, Not tested.

IFA: anti-idiotype-sensitized mice showed antigen-specific DTH reactions, and their spleen cells showed increased proliferative responses to stimulation with the purified 38-kDa antigen (p38).

Surprisingly, injection of Rb71 in IFA did not result in an idiotype-positive serum immunoglobulin response such as was observed following immunization with Rb04 or Rb10 (Table 2). These empirical results, at least with the examined anti-idiotypes, indicate a reciprocal relationship between the T-cell and B-cell (Id$^+$)-stimulatory internal image. The lack of correlation between these respective functions is not surprising, since it has also been reported for peptides of certain structural orientation (10). Immunogenetic factors could be important, and their analysis may illuminate this subject, particularly since the antibody response to the TB71 epitope of the p38 antigen has been found to be under H-2 (IA) control (21).

The antigen-specific stimulation of T cells with an anti-idiotype image, produced against the private TB71 idiotype, raises two problems of interpretation. First, it is difficult to reconcile the idea that antibodies directed to a determinant of the antigen-combining site of immunoglobulin would have a structural homology with the T-cell receptor. The absence of V-gene rearrangement in T cells argues against the notion that the same genes that encode immunoglobulin V regions are functional in T-cell receptors (27), although it has also been suggested that T cells do possess a receptor with segmental homology to the V region of immunoglobulins (31). If one postulated a steric rather than sequential homology between the immunoglobulin and T-cell recognition site, the occurrence of the overlapping epitope specificity would still be exceptional, since these are mutually exclusive in most protein antigens. The exceptional occurrence is supported by the fact that the T-stimulatory potential was found in only one of several tested

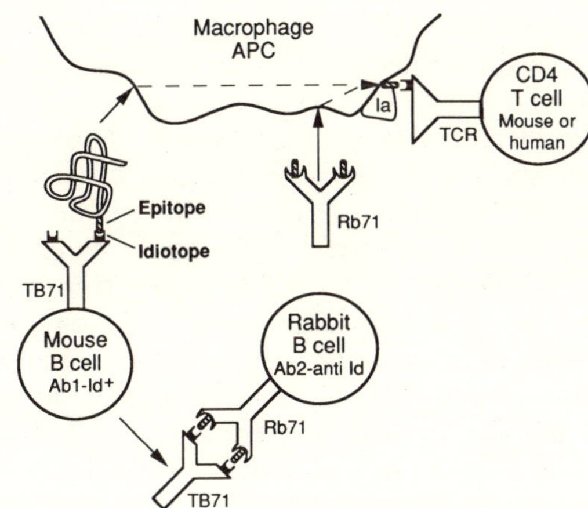

Figure 1. Schematic representation of the T-cell stimulatory internal-image idiotype.

anti-idiotypic antibodies. One may speculate that antigen processing by TB71-specific B cells could have protected the epitope for presentation to T cells of corresponding specificity. This mechanism, schematically represented in Fig. 1, may provide the basis for the overlapping recognition by antibody and T cells.

The second unexpected feature of the phenomenon is the disparity between the frequent, i.e., public occurrence of the idiotype on T cells (detected by proliferative or DTH assay) and the apparently private incidence of B cells which secrete Rb71-positive antibodies. Although there is no fundamental obstacle to such a situation, further studies are needed to justify the proposed explanation. Recent advances in establishing the full sequence of the p38 antigen (1) should enable a more systematic analysis of the integral epitope structure and of the anti-idiotypic image of epitopes of this antigen.

ANTI-IDIOTYPIC STIMULATION OF HUMAN T CELLS

The rabbit anti-idiotypic antibody RB71 can stimulate in vitro proliferation of human peripheral blood lymphocytes (45). Of eight individuals tested, four were responders and four were nonresponders to RB71, and the anti-idiotypic responsiveness was strictly linked to that to the corresponding p38 antigen. Three of the anti-idiotypic responders were healthy, BCG-vaccinated individuals, indicating that T-cell responsiveness, unlike serum antibody levels to this antigen (18, 24), does not discriminate between sensitized healthy individuals and patients with tuberculosis. In view of the 30% structural homology of p38 with the *phoS* periplasmic protein of *Escherichia coli* (1), responder individuals could also have been sensitized by contact with commensal organisms containing a T-cell cross-reactive homologous protein. Individual variation in responsiveness to Rb71 and p38 was not fully overlapping in a study of 30 healthy donors (44) and was not significantly related to the human leukocyte antigen DR2 haplotype, which was shown to be associated with elevated antibody levels to the corresponding TB71 epitope in patients with pulmonary tuberculosis (6).

The mechanisms of stimulation by RB71 have been investigated with a T-cell clone (CW4.13) selected by stimulation with the p38 antigen (45, 46). However, three other anti-p38 clones did not respond to RB71, indicating an incomplete overlap in specificity between the T-cell-stimulatory domain of RB71 and the native p38 antigen, which apparently contains additional epitopes not represented in RB71. The proliferative response of CW4.13 to RB71 as well as to p38 was dependent on the presence of accessory cells in culture. Autologous irradiated peripheral blood mononuclear cells and Epstein-Barr virus-transformed lymphoblastoid cells were of comparable activity. Moreover, the proliferative response of purified blood lymphocytes was also accessory-cell dependent, indicating that this feature is not unique to the stimulation of the CW4.13 clone. Since the proliferation of CW4.13 cells required human leukocyte antigen DR2-matched accessory cells, the data suggest that the stimulatory domain of RB71 was presented in association with class II major histocompatibility complex molecules, rather than by cross-linking to Fc receptors, as for the stimulation with anti-CD3 antibodies. The latter mechanism was also excluded by the failure of Fc receptor blockade with normal rabbit globulin to prevent the RB71 stimulation. Furthermore, the lack of stimulation with Sepharose-linked RB71 also suggested a mechanism distinct from the anti-CD3 receptor-mediated response. The most compelling evidence suggesting that the intact combining site of the anti-idiotype is not required for T-cell stimulation came from experiments with reduced and S-carboxymethylated affinity-purified RB71 (46). This preparation with ablated idiotype-binding activity, containing dissociated H and L chains, produced a response of T cells comparable with that of intact RB71 immunoglobulin, thus suggesting that stimulation was not dependent on the conformational integrity of the anti-idiotype combining site. Attempts to localize the stimulatory activity between separated chains, however, produced variable results between individual cell donors (46). When taken together, these results suggest that the internal-image domain of anti-idiotype Rb71 is presented to specific T cells by similar mechanisms to the determinants of protein antigens in general. Presumably, the anti-idiotype after processing would have a critical peptide of its combining site reexpressed in association with class II major histocompatibility complex determinants. Although the specificity of T-cell epitopes is determined primarily by the amino acid structure, it is not implausible that the anti-idiotype internal image is a highly selected structure with mimicry based on alternative residues.

AUTO-ANTI-IDIOTYPIC ANTIBODIES IN T-CELL ANERGY

The immunological defect of lepromatous leprosy is characterized by a lack of specific T-cell immunity to *M. leprae.* A certain proportion of patients fail to respond by DTH to *M. leprae* extracts while responding to tuberculin (purified protein derivative [PPD]), thus indicating a suppression of response to cross-reactive antigens by a reaction to *M. leprae*-specific determinants (3). An experimental model with at least some similar features is represented by DTH anergy and nonspecific suppression of immune reactions by spleen cells from heavily BCG-infected mice (7, 58). The suppression in infected mice was attributed to activated macrophages, whereas the possible role of T cells remains equivocal (37).

The role of idiotype-specific antibodies was proposed on the basis of experiments in which the putative auto-anti-idiotypic antibodies from PPD-anergic mice infected intravenously with 2×10^7 BCG organisms blocked the adoptive cell transfer of DTH to PPD (9). The targeting of this suppressive activity was antigen specific since it did not apply to contact sensitivity to oxazolone and was mediated by antibodies directed to anti-PPD idiotype rather than PPD (antigen), as determined by cross-absorption analysis with coupled Sepharose columns. The DTH-suppressive potency of serum was most pronounced 21 to 28 days postinfection but negligible at both 7 and 60 days after BCG infection. This distinct time course supports the authors' opinion that complexes between antibodies and persisting BCG could represent a particularly good stimulus for the production of anti-idiotypic antibodies. The described phenomenon implies an overlapping specificity between antibody and T-cell idiotypes, analogous to the situation with the Rb71 idiotype but with blocking rather than stimulation of DTH reaction. As an alternative to direct blocking of DTH effector cells, Colizzi and his colleagues (9) proposed that anti-idiotypic antibodies acting indirectly could trigger suppressor cells which, in turn, would inhibit the effector stage of DTH. The diverse biological effects of anti-idiotypes with respect to the T-cell response may reflect the distinct idiotypic and consequently the antigenic specificities of functionally diverse T-cell subsets. However, since PPD contains several mycobacterial antigens, mostly in a degraded form, the molecular specificity of the auto-anti-idiotype serum requires further analysis.

A hypothesis was formulated about the possible role of auto-anti-idiotypic antibodies for the immunopathogenesis of lepromatous leprosy; it proposed that lepromatous leprosy may develop as a result of anti-idiotypic-mediated chronic suppression of T-cell immunity (14). By analogy with results in mouse models, it was argued that antigen responsiveness could have been altered (suppressed) by perinatal exposure to idiotype-bearing anti-*M. leprae* or anti-idiotype antibodies of maternal origin. Although total antibody levels could be compensated for by clones of alternative idiotypic specificity, it was proposed that for T cells a more restricted idiotype repertoire would enable effective suppression to occur. It can further be speculated that an idiotypic specificity with even a private antibody profile but public T-cell expression could indeed satisfy the requirements for the expected T-cell suppression. Other aspects of this hypothesis related to genetic factors, particularly to the independence of inheritance of idiotypes from MHC genes (39, 51), as a possible basis for the development of leprosy in only a small proportion of infected individuals. Although idiotype-controlling genes in conjunction with other genetic factors may determine cases of a priori lepromatous disease, the effects of exposure to environmental antigens on idiotype expression may play a role in downgrading from tuberculoid to lepromatous forms of the disease spectrum. Mycobacterial antigens which are antigenically homologous with the highly conserved stress proteins represented in several commensal organisms (60) seem to be the most likely candidates for the latter mechanism.

PUBLIC IDIOTYPES

There is ample evidence that sera of patients with leprosy and tuberculosis contain a variety of autoantibodies reacting with DNA and other polynucleotides, nuclear antigen, cardiolipin, thyroglobulin, rheumatoid factor, or germinal cells (30, 49, 62). The idiotype markers of these antibodies have been studied in several laboratories to find whether they represent the expression of a limited repertoire associated with the mycobacterial infection or a random polyclonal stimulation. Generally, these idiotypes can be classified as cross-reactive for antibodies of several specificities.

Sera of patients with tuberculosis have been examined for the presence of the 16/6 idiotype (49). This marker is carried by a human monoclonal immunoglobulin M anti-DNA antibody from a patient with systemic lupus erythematosus and is detected by rabbit or mouse monoclonal anti-idiotype reagents. Sixty percent (35 of 57 tested) of sera from patients with active pulmonary tuberculosis were found to be 16/6 idiotype positive, compared with only 1% positivity in the matched control group. These results suggested that autoantibodies in patients with tuberculosis and those in patients with systemic lupus erythematosus, in whom the 16/6 idiotype is commonly found, are generated by a corresponding B-cell network and possibly by similar mechanisms. Since 16/6 idiotype levels are correlated with serum immunoglobulin G and immunoglobulin M hypergammaglobulinemia, it has been argued that the increased autoantibody levels result from the polyclonal activation of B cells by the known adjuvant effects of mycobacterial constituents. This view is supported by the finding that polyclonal activation of lymphocytes from healthy individuals with several bacterial polyclonal activators, including *Klebsiella* spp., *Staphylococcus* spp., *Streptococcus* spp., and Epstein-Barr virus, stimulate 16/6 idiotype-positive autoantibody formation (13, 49). However, there also seems to be a selective element, since a common idiotype associated with autoantibodies to hepatitis B surface antigen was not found to be elevated in patients with tuberculosis.

Several other idiotypes of human MAbs with specificity to DNA, mitochondria, or acetylcholine receptor were identified (30). MAbs derived from either leprosy or systemic lupus erythematosus patients displayed cross-reactive idiotypes, which cross-reacted mutually and also with the 16/6 antigen. However, two rabbit antisera, directed to 8E7 and TH9 idiotypes, seemed to identify somewhat distinct specificities. Since these reagents recognized antibodies in individuals of widely different ethnic and geographic origins, the immunoglobulins bearing the detected common idiotypes could have been conserved in the germ line repertoire. Two other common idiotypes, designated PR4 and TH3, carried by human MAbs originating from patients with leprosy, were identified (28, 62). Both antibodies reacted with a determinant carried by the phenolic glycolipid of *M. leprae* (distinct from its specific terminal disaccharide epitope), single- and double-stranded DNA, poly(ADP-ribose), and poly(dT); PR4 also stained the cytoplasm of basal keratinocytes of interfollicular epidermis. Cross-competition assays indicated different but related idiotype specificities: PR4 was similar to and TH3 was distinct from the 16/6 idiotype. It is of interest that binding of both anti-idiotype reagents to their respective idiotypes was inhibited by 1 to 10 µg of antigen per ml, indicating the localization of the idiotypic determinant within or close to the antibody-combining site.

The distribution of the PR4 idiotype was examined by using sera from both leprosy and tuberculosis patients (61). Positivity was found in about 60% of lepromatous leprosy patients, 50% of tuberculoid leprosy patients, and 37% of pulmonary tuberculosis patients; it was unrelated to immunoglobulin levels in serum. These results were found irrespective of exposure to malaria infection, for which polyclonal B-cell activation and autoantibody formation have been described. On the other hand, in Sjögren's syndrome, which is also associated with hypergammaglobulinemia, only 15% of sera were PR4 positive (59). In view of the somewhat different results for PR4 than for 16/6, it appears that distinct idiotypically defined germline products may have individual patterns of expression rather than being a uniform outcome of polyclonal B-cell stimulation. Since PR4 binds to a determinant which is shared between mycobacterial and host cell constituents, the authors (61) suggested that its persistence may result from foiled B-cell somatic mutation by the presence of excessive antigenic stimulation.

REFERENCES

1. **Andersen, A. B., and E. B. Hansen.** 1989. Structure and mapping of antigenic domains of protein antigen b, a 38,000-molecular-weight protein of *Mycobacterium tuberculosis. Infect. Immun.* **57:**2481–2488.

2. **Auchincloss, H., A. Jeffrey, Jr., A. Bluestone, and D. H. Sachs.** 1983. Anti-idiotypes against anti-H2 monoclonal antibodies. V. *In vivo* anti-idiotype treatment induces idiotype-specific helper T cells. *J. Exp. Med.* **157:**1273–1286.

3. **Bloom, B. R., and V. Mehra.** 1984. Immunological unresponsiveness in leprosy. *Immunol. Rev.* **80:**5–28.

4. **Bluestone, J. A., L. Oberdan, S. L. Epstein, and D. H. Sachs.** 1986. Idiotypic manipulation of the immune response to transplantation antigens. *Immunol. Rev.* **90:**5–27.

5. **Bothamley, G., P. Udani, R. Rudd, F. Festenstein, and J. Ivanyi.** 1988. Humoral response to defined epitopes of tubercle bacilli in adult pulmonary and child tuberculosis. *Eur. J. Clin. Microbiol. Infect. Dis.* **7:**639–645.

6. **Bothamley, G. H., J. S. Beck, G. M. Schreuder, T. J. D'Amaro, R. R. P. de Vries, T. Kardjito, and J. Ivanyi.** 1989. Association of tuberculosis and *M. tuberculosis*-specific antibody levels with HLA. *J. Infect. Dis.* **159:**549–555.

7. **Brown, C. A., and I. N. Brown.** 1982. *Mycobacterium bovis,* BCG, modulation of murine antibody responses: influence of dose and degree of aggregation of live or dead organisms. *Br. J. Exp. Pathol.* **63:**133–143.

8. **Coates, A. R. M., J. Hewitt, B. W. Allen, J. Ivanyi, and D. A. Mitchison.** 1981. Antigenic diversity of *Mycobacterium tuberculosis* and *Mycobacterium bovis* detected by means of monoclonal antibodies. *Lancet* **ii:**167–169.

9. **Colizzi, V., M. Giuntini, C. Garzelli, M. Campa, and G. Falcone.** 1983. Auto-anti-idiotypic antibodies inhibit T-cell-mediated hypersensitivity in BCG-infected mice. *Cell. Immunol.* **80:**205–210.

10. **Cox, J. H., J. Ivanyi, D. B. Young, J. R. Lamb, A. D. Syred, and M. J. Francis.** 1988. Orientation of epitopes influences the immunogenicity of synthetic peptide dimers. *Eur. J. Immunol.* **18:**2015–2019.

11. **Eichmann, K., F. Emmrich, and S. H. E. Kaufmann.** 1987. Idiotypic vaccination: considerations towards a practical application. *Crit. Rev. Immunol.* **7:**192–227.

12. **Eichmann, K., and K. Rajewsky.** 1975. Induction of T and B cell immunity by anti-idiotypic antibody. *Eur. J. Immunol.* **5:**661–666.

13. **El-Roeiy, A., W. L. Gross, J. Ludemann, D. A. Isenberg, and Y. Shoenfeld.** 1986. Preferential secretion of a common anti-DNA idiotype (16/6 Id) and anti-polynucleotide antibodies by normal mononuclear cells following stimulation with Klebsiella pneumoniae. *Immunol. Lett.* **12:**313–319.

14. **Ferluga, J., V. Colizzi, A. Ferrante, M. J. Colston, and E. J. Holborow.** 1984. Hypothesis: possible idiotypic suppression of cell-mediated immunity in lepromatous leprosy. *Lepr. Rev.* **55:**221–227.

15. **Francotte, M., and J. Urbain.** 1985. Enhancement of antibody response by mouse dendritic cells pulsed with tobacco mosaic virus or with rabbit anti-idiotypic antibodies raised against a private rabbit idiotype. *Proc. Natl. Acad. Sci. USA* **82:**8149–8152.

16. **Hart, D. A., A. L. Wang, L. L. Pawlak, and A. Nisonoff.** 1972. Suppression of idiotypic

specificities in adult mice by administration of anti-idiotypic antibody. *J. Exp. Med.* **135**:1293–1300.

17. **Ivanyi, J.** 1986. Pathogenic and protective interactions in mycobacterial infections, p. 127–157. *In* V. Britten and H. P. A. Hughes (ed.), *Immunological Recognition of Altered Cell Surfaces in Infection and Disease. Clinics in Immunology,* vol. 6. The W. B. Saunders Co., Philadelphia.

18. **Ivanyi, J., G. H. Bothamley, and P. S. Jackett.** 1988. Immunodiagnostic assays for tuberculosis and leprosy. *Br. Med. Bull.* **4**:635–649.

19. **Ivanyi, J., E. Krambovitis, and M. Keen.** 1983. Evaluation of a monoclonal antibody (TB72) based serological test for tuberculosis. *Clin. Exp. Immunol.* **54**:337–345.

20. **Ivanyi, J., J. A. Morris, and M. Keen.** 1985. Studies with monoclonal antibodies to mycobacteria, p. 59–90. *In* A. J. L. Macario and E. C. Macario (ed.), *Monoclonal Antibodies against Bacteria,* vol. 1. Academic Press, Inc., New York.

21. **Ivanyi, J., and K. Sharp.** 1986. Control by H-2 genes of murine antibody responses to protein antigens of *Mycobacterium tuberculosis. Immunology* **59**:329–332.

22. **Ivanyi, J., K. Sharp, P. Jackett, and G. Bothamley.** 1988. Immunological study of the defined constituents of mycobacteria. *Springer Semin. Immunopathol.* **10**:279–300.

23. **Ivanyi, J., S. Sinha, R. Aston, D. Cussel, M. Keen, and U. Sengupta.** 1983. Definition of species specific and cross-reactive antigenic determinants of *Mycobacterium leprae* using monoclonal antibodies. *Clin. Exp. Immunol.* **52**:528–536.

24. **Jackett, P. S., G. H. Bothamley, H. V. Batra, A. Mistry, D. B. Young, and J. Ivanyi.** 1988. Specificity of antibodies to immunodominant mycobacterial antigens in pulmonary tuberculosis. *J. Clin. Microbiol.* **26**:2313–2318.

25. **Kaufmann, S. H. E., and I. E. A. Flesh.** 1988. The role of T cell-macrophage interactions in tuberculosis. *Springer Semin. Immunopathol.* **10**:337–358.

26. **Kelsoe, G., M. Reth, and K. Rajewsky.** 1980. Control of idiotype expression by monoclonal anti-idiotype antibodies. *Immunol. Rev.* **52**:75–88.

27. **Kronenberg, M., E. Kraig, G. Siu, J. A. Kapp, J. Kappler, P. Marack, C. W. Pierce, and L. Hood.** 1983. Three T-cell hybridomas do not contain detectable heavy chain variable region gene transcripts. *J. Exp. Med.* **158**:210–227.

28. **Lockniskar, M., A. Zumla, D. Mudd, D. A. Isenberg, W. Williams, and K. P. W. J. McAdam.** 1988. Human monoclonal antibodies to phenolic glycolipid-I derived from patients with leprosy, and production of specific anti-idiotypes. *Immunology* **64**:245–251.

29. **Lovik, M., and O. Closs.** 1982. Repeated delayed-type hypersensitivity reactions against *Mycobacterium lepraemurium* antigens at the infection site do not affect bacillary multiplication in C3H mice. *Infect. Immun.* **36**:768–774.

30. **Mackworth-Young, C., J. Sabbaga, and R. S. Schwartz.** 1987. Idiotypic markers of polyclonal B cell activation: public idiotypes shared by monoclonal antibodies derived from patients with systemic lupus erythematosus or leprosy. *J. Clin. Invest.* **79**:572–581.

31. **Marchalonis, J. J.** 1985. Antigen-binding molecules of T cells: distinction from MHC-restricted molecules and segmental homology to immunoglobulin V_H and T-cell receptor genes. *Scand. J. Immunol.* **21**:99–107.

32. **Mehra, V., D. Sweetser, and R. A. Young.** 1986. Efficient mapping of protein antigenic determinants. *Proc. Natl. Acad. Sci. USA* **83**:7013–7017.

33. **Mills, K. H. G., J. J. Skehel, and D. B. Thomas.** 1986. Conformational-dependent recognition of influenza virus hemagglutinin by murine T helper clones. *Eur. J. Immunol.* **16**:276–280.

34. **Modlin, R. L., and T. H. Rea.** 1988. Immunopathology of leprosy granulomas. *Springer Semin. Immunopathol.* **10**:359–374.

35. **Moore, M. J., and J. Ivanyi.** 1989. Idiotype vaccines. *In* E. Liew (ed.), *Vaccines in Tropical Diseases.* CRC Press, Inc., Boca Raton, Fla., in press.

36. **Noguiera, N., G. Kaplan, E. Levy, E. N. Sarno, P. Kushner, A. Granelli-Piperno, L. Vieira, V. Colomer Gould, W. Levis, R. Steinman, Y. K. Yipl, and Z. A. Cohn.** 1983. Defective interferon production in leprosy. Reversal with antigen and interleukin 2. *J. Exp. Med.* **158:**2165–2170.

37. **Orme, I. M., and F. M. Collins.** 1984. Immune response to atypical mycobacteria: immunocompetence of heavily infected mice measured in vivo fails to substantiate immunosuppression data obtained in vitro. *Infect. Immun.* **43:**32–37.

38. **Oudin, J., and P. A. Cazenave.** 1971. Similar idiotypic specificities in immunoglobulin fractions with different antibody functions or even without detectable antibody function. *Proc. Natl. Acad. Sci. USA* **68:**2616–2621.

39. **Pasquali, J. L., S. Fong, C. Tsoukes, J. H. Vaughan, and D. A. Carson.** 1980. Inheritance of immunoglobulin M rheumatoid-factor idiotypes. *J. Clin. Invest.* **66:**863–866.

40. **Praputpittaya, K.** 1986. Anti-idiotypic antibodies and their role in the immune response to *M. leprae* and *M. tuberculosis.* Ph.D. thesis. University of Birmingham, Birmingham, United Kingdom.

41. **Praputpittaya, K., and J. Ivanyi.** 1987. Study of idiotypes expressed by monoclonal antibodies to the 35 kD and 12 kD antigens of *Mycobacterium leprae. Clin. Exp. Immunol.* **70:**298–306.

42. **Praputpittaya, K., and J. Ivanyi.** 1987. Stimulation by anti-idiotype antibody of murine T cell responses to the 38 kD antigen of *Mycobacterium tuberculosis. Clin. Exp. Immunol.* **70:**307–315.

43. **Prasad, H. K., R. S. Mishra, and I. Nath.** 1987. Phenolic glycolipid-I of *Mycobacterium leprae* induces general suppression of *in vitro* concanavalin A responses unrelated to leprosy type. *J. Exp. Med.* **165:**239–244.

44. **Rees, A. D. M., and J. R. Lamb.** 1989. Human T cell recognition of an "internal image" of a mycobacterial antigen in an anti-idiotypic antibody. *Proceedings: Idiotype Networks in Biology and Medicine.* Elsevier Science Publishing, Inc., New York, in press.

45. **Rees, A. D. M., K. Praputpittaya, A. Scoging, N. Dobson, J. Ivanyi, D. Young, and J. R. Lamb.** 1987. T cell activation by anti-idiotypic antibody: evidence for the internal image. *Immunology* **60:**389–393.

46. **Rees, A. D. M., A. Scoging, N. Dobson, K. Praputpittya, D. Young, J. Ivanyi, and J. R. Lamb.** 1987. T cell activation by anti-idiotypic antibody: mechanism of interaction with antigen-reactive T cells. *Eur. J. Immunol.* **17:**197–201.

47. **Rook, G. A. W., J. W. Carswell, and J. L. Stanford.** 1976. Preliminary evidence for the trapping of antigen-specific lymphocytes in the lymphoid tissue of 'anergic' tuberculosis patients. *Clin. Exp. Immunol.* **26:**129–132.

48. **Roth, C., G. Somme, M. L. Gougeon, and J. Theze.** 1985. Induction by monoclonal anti-idiotypic antibodies of an anti-poly-(Glu^{60} Ala^{30} Tyr^{10}) (GAT) immune response in GAT responder and non-responder mice. *Scand. J. Immunol.* **21:**361–367.

49. **Sela, O., A. El-Roeiy, D. A. Isenberg, R. C. Kennedy, C. B. Colaco, J. Pinkhas, and Y. Shoenfeld.** 1987. A common anti-DNA idiotype in sera of patients with active pulmonary tuberculosis. *Arthritis Rheum.* **30:**50–57.

50. **Sengupta, U., S. Sinha, G. Ramu, J. Lamb, and J. Ivanyi.** 1987. Suppression of delayed hypersensitivity skin reactions to tuberculin by *M. leprae* antigens in patients with lepromatous and tuberculoid leprosy. *Clin. Exp. Immunol.* **68:**58–64.

51. **Sercarz, E., L. Wicker, J. Stratton, A. Miller, D. Metzger, R. Maizels, M. Katz, M. Harvey, and C. Benjamin.** 1981. Regulation of antibody specificity and idiotype by two independent T cells, p. 533–546. *In* C. Janeway, E. E. Sercarz, and H. Wigzell (ed.), *Immunoglobulin Idiotypes.* Academic Press, Inc., New York.

52. **Sharpe, A. H., G. N. Gaulton, K. K. McDade, B. N. Fields, and M. I. Green.** 1984.

Syngeneic monoclonal anti-idiotype can induce cellular immunity to reovirus. *J. Exp. Med.* **160**:1195–1205.

53. **Sy, M.-S., A. R. Brown, B. Benacerraf, and M. I. Green.** 1980. Antigen and receptor-driven regulatory mechanisms. III. Induction of delayed-type hypersensitivity to azobenzenearsonate with anti-cross-reactive idiotypic antibodies. *J. Exp. Med.* **151**:896–909.

54. **Takemori, T., H. Tesch, M. Reth, and K. Rajewsky.** 1982. The immune response against anti-idiotype antibodies. I. Induction of idiotope-bearing antibodies and analysis of the idiotope repertoire. *Eur. J. Immunol.* **12**:1040–1046.

55. **Thomas, W. R., G. Morahan, I. D. Walker, and J. F. A. P. Miller.** 1981. Induction of delayed-type hypersensitivity to azobenzenearsonate by a monoclonal anti-idiotype antibody. *J. Exp. Med.* **153**:743–747.

56. **Turk, J. L., and D. Parker.** 1982. Effect of cyclophosphamide on immunological control mechanisms. *Immunol. Rev.* **65**:99–113.

57. **Uytdehaag, F., and O. Osterhaus.** 1985. Induction of neutralizing antibody in mice against poliovirus type II with monoclonal anti-idiotypic antibodies. *J. Immunol.* **134**:1225–1229.

58. **Watson, S. R., and F. M. Collins.** 1981. The specificity of suppressor T cells induced by chronic *Mycobacterium avium* infection in mice. *Clin. Exp. Immunol.* **43**:10–19.

59. **Williams, W., A. Zumla, R. Behrens, M. Locniskar, A. Voller, K. P. W. J. McAdam, and D. A. Isenberg.** 1988. Studies of a common idiotype PR4 in autoimmune rheumatic disease. *Arthritis Rheum.* **31**:1097–1104.

60. **Young, D. B.** 1988. Structure of mycobacterial antigens. *Br. Med. Bull.* **44**:562–583.

61. **Zumla, A., W. Williams, D. Mudd, M., Locniskar, R. Behrens, D. A. Isenberg, and K. P. W. J. McAdam.** 1989. Identification of a common idiotype PR4 in patients with leprosy and tuberculosis. Submitted for publication.

62. **Zumla, A., W. Williams, S. Shall, M. Locniskar, I. Leigh, K. P. W. J. McAdam, and D. A. Isenberg.** 1988. Human monoclonal antibodies to phenolic glycolipid-1 from leprosy patients cross-react with poly(ADP-ribose), polynucleotides and tissue bound antigens. *Autoimmunity* **1**:183–195.

Chapter 7

The Idiotypic Network and Immediate Hypersensitivity

Jean-Marie R. Saint-Remy

Allergic diseases are frequent: 10 to 15% of the population of Western Europe and the United States will, at some time during their lifetime, suffer from allergy-related symptoms. It is therefore not surprising that allergic diseases rank second among the main causes of absenteeism from work or school. The spectrum of clinical syndromes related to allergy is wide, from the benign but bothersome hay fever to death from anaphylactic shock due to, for instance, penicillin hypersensitivity.

Allergy, or immediate hypersensitivity, depends on the presence of antibodies of the immunoglobulin E (IgE) isotype. These antibodies bind to high-affinity specific receptors on the surface of basophils and mastocytes through amino acid residues located in the third and fourth domains of the ε heavy chain. The bridging of at least two cell-bound IgE antibodies by the corresponding allergen leads to activation of a membrane phospholipase and an influx of calcium. The cell degranulates, freeing preformed histamine, and leukotrienes, prostaglandins, and platelet-activating factor are synthesized through activation of the arachidonic acid metabolic pathways. These mediators are responsible for the development of allergic symptoms.

The allergen bridging of specific IgE molecules located on mastocyte surfaces plays a central role in these events, and it may therefore be assumed that attempts to modulate allergen binding to IgE molecules could represent a valuable clinical approach to the treatment of allergic diseases. More fundamentally, of course, modulation of the IgE immune response itself could be of great clinical benefit.

I will (i) review briefly how a significant reduction of allergen-IgE

Jean-Marie R. Saint-Remy • Experimental Medicine Unit, Institute of Cellular and Molecular Pathology, Université Catholique de Louvain, 1200 Brussels, Belgium.

interaction at the surface of basophils or mast cells can be achieved; (ii) consider the theoretical potentials and hazards of idiotypic manipulation of the IgE immune response; (iii) examine some characteristics of the IgE immune response which indicate that IgE regulation may be exquisitely sensitive to idiotypic regulation; (iv) review work carried out on idiotypic suppression of IgE production in animals; and (v) consider recent data gathered in human immediate hypersensitivity.

ALTERATION OF THE INTERACTIONS BETWEEN IgE ANTIBODIES, ALLERGENS, AND MASTOCYTES

The first obvious way in which binding of allergens to cell-bound IgE can be controlled is by reducing allergen exposure. Avoidance of allergens over long periods has resulted not only in a disappearance of acute-symptom exacerbations but also in an improvement of nonspecific reactivity regularly associated with immediate hypersensitivity, as, for instance, in bronchial asthma (91). Allergen exposure should therefore be reduced whenever possible, but it should be realized that allergen avoidance measures often remain unrealistic.

A second means by which a reduction of allergen binding to cell-bound IgE antibodies can be achieved is through competition for allergen binding by other isotypes. The rationale for conventional hyposensitization, a treatment used for more than 70 years and involving regular inoculations of the sensitizing allergen (29), has been the induction of a relative increase in specific antiallergen antibodies of isotypes unable to bind to basophils or mast cells. This therapy has been successful in systemic hypersensitivity to Hymenoptera insect venoms and in spasmodic rhinitis, but its usefulness in allergic asthma remains questionable (38). Moreover, in many cases hyposensitization cannot be performed, either because the precise nature of the allergen is unknown (e.g., in food allergy) or because the treatment is too hazardous (e.g., in drug hypersensitivity). Side effects of hyposensitization are frequent but often minor; the occurrence of fatalities has nevertheless led health authorities in certain countries to prohibit its use unless full resuscitation facilities are at hand. The precise mode of action of conventional hyposensitization has not been elucidated but, inter alia, there have been suggestions of induction of anti-idiotypic antibodies (41), as will be discussed below.

If the binding of IgE antibodies to the IgE-specific cell receptor could be inhibited, mediator release and, hence, symptoms of allergy would also be reduced. Synthetic peptides representative of cytophilic domains of IgE antibodies have been used to this end in both animal and clinical studies (58). Although it is too early to decide whether these peptides will be useful in particular clinical situations, their rapid metabolism and relatively weak affinity for cell receptors, as compared with the whole IgE molecule, impede their use as a treatment of immediate hypersensitivity at present.

The regulation of IgE antibody production at the isotype level has recently received much attention. IgE production results from the differentiation of precommitted B cells through complex interactions between cells and soluble cytokines, namely, interleukin-4 (IL-4), IL-5, and IL-6 (27, 115), and is antagonized by gamma interferon (106). It has been speculated that such a detailed understanding of the mechanisms by which IgE antibodies are produced will yield therapeutic applications. Each of these lymphokines, however, has a multitude of functions in addition to IgE regulation; some of them are related to tissue sensitivity: IL-3 is a mast cell stimulation factor (32), and IL-5, besides its synergic effect with IL-4 on IgE production (90), acts as a selective eosinophil activator (71). This multiplicity of functions makes it difficult to predict the outcome of therapies whose aim would be the control of production or function of lymphokines. Moreover, it is not clear whether an isotypic suppression of IgE antibody production has detrimental effects. IgE antibodies are essential in host defense against parasitic infections (19, 65) and, presumably, against other infectious diseases (104).

Strategies for antigen-specific regulation of IgE antibody production can be divided into two main categories. Chemical modification of the allergen itself can generate derivatives which induce a specific suppression of IgE antibody formation, as with derivatives of monomethoxypolyethylene glycol (120). However, up to now, experimental data have shown significant variations in the physiological outcome of inoculation of these derivatives, depending on the model used. Identification by protein sequencing of relevant allergen epitopes (26) will no doubt offer new perspectives in this area.

Antigen-specific IgE regulation can be exerted at the level of the antibody itself, i.e., its idiotype. In theory, this approach would result in a safe (since no free allergen is used) and selective regulation of IgE. However, it should be realized that there exists a great diversity of antibodies, and hence of idiotypes, in hypersensitive patients, since they are frequently sensitized to several allergens which are macromolecules bearing a number of epitopes. I will demonstrate, by referring to animal experiments and recent clinical experience, that, in practice, these conceptual difficulties can be surmounted.

IDIOTYPY AND IMMEDIATE HYPERSENSITIVITY

I shall first consider how the present understanding of idiotypic regulation of antibody production might be applied to the control of immediate hypersensitivity and then discuss potential hazards of therapies based on such regulation.

Potential of Idiotypic Manipulation in Immediate Hypersensitivity

Some anti-idiotypic antibodies which are raised to the paratope—the antigen-binding site—of specific anti-antigen antibodies are thought to carry an internal image of the antigen. These are able to inhibit the binding of specific antibodies to antigen. Anti-idiotypic antibodies in this category ($Ab_2\beta$ [67]) might be used as substitutes for allergens to increase the specific IgG immune response. The advantages of this alternative to conventional hyposensitization include a better characterization of the material used for inoculation and, therefore, of the induced immune response. The dangers inherent in whole-allergen injection would also be reduced, especially since in humans, specific IgE and IgG antibodies are probably directed toward different epitopes (97). Although much progress has been made in the application of this approach in infectious diseases (95), no definitive attempt to apply this concept to immediate hypersensitivity has yet been published. Its practical interest might, however, be restricted by the recent success obtained in the sequencing of main allergens; identification of relevant allergen epitopes could permit the synthesis of derivatives either chemically or by molecular biology.

Anti-idiotypic antibodies, when administered in the nanogram concentration range, are known to be potent in vivo enhancers of an immune response (33), increasing the frequency of Id^+ precursors (35). It is therefore conceivable that the production of specific IgG antibodies is increased either by passive transfer or by active production of anti-idiotypic antibodies. However, the precise mechanism by which enhancement of the immune response is produced is unknown, but probably depends on the isotype of anti-idiotypic antibodies and on their complement activation properties (68). Moreover, differences between animal species have been demonstrated (92). No deliberate effort to use anti-idiotypic antibodies for this purpose in human immediate hypersensitivity has been reported, but it may be speculated that the clinical efficiency of passive administration of selected pooled immunoglobulins (1) is at least partly due to the presence of small amounts of anti-idiotypic antibodies.

Finally, selective suppression of antibody production by anti-idiotypic antibodies has been achieved both in vitro (12) and in vivo (34). In vivo, both passive transfer of heterologous or isologous anti-idiotypic antibodies and active production of anti-idiotypic antibodies have been used successfully. Convincing evidence of idiotype-related suppression of IgE antibody production has been published, indicating that IgE might be particularly sensitive to such regulation (5, 79). Applications of such protocols to human hypersensitivity might provide a highly specific way to reduce the production of IgE antibodies to a given allergen. Interestingly, in diseases such as thrombocytopenia or hemophilia A with anti-factor VIII antibodies, repeated infusions of immunoglobulins result in a significant improvement; again, part of this efficiency may depend on the presence of anti-idiotypic antibodies in pooled immunoglobulins, as recently claimed (94).

In an attempt to analyze the feasibility of idiotype suppression in human immediate hypersensitivity, we have characterized the human immune response to allergens of *Dermatophagoides pteronyssinus* (the common house dust mite) and to grass pollens. Results of these in vitro studies have led to a first clinical application, which has raised some hopes that this might be developed into a new therapy for immediate hypersensitivity. This work is summarized in Clinical Trials (see below).

Possible Hazards of Idiotypic Suppression in Immediate Hypersensitivity

Idiotypic suppression of specific IgE antibody production might have detrimental consequences for patients. Anti-idiotypic antibodies specific to IgE antibodies could provoke the degranulation of mastocytes or basophils. This was shown in vitro when rabbit anti-idiotypic antibodies raised to a monoclonal antibody specific for one of the major allergens of Rye grass, *Lol pI,* and cross-reacting with human IgE antibodies to *Lol pI,* induced the release of histamine when they were added to basophils of *Lol pI*-sensitized patients (82). In vivo, a passive cutaneous anaphylactic (PCA) reaction can be induced by isologous anti-idiotypic antibodies in rats sensitized with mouse anti-benzylpenicilloyl (BPO) IgE antibodies (87). Rabbit anti-idiotypic antibodies raised to anti-tetanus toxoid (TT) human IgG antibodies elicit a typical wheal-and-flare reaction after inoculation into normal human skin previously sensitized with a serum containing anti-TT IgE antibodies (46). Taken together, these observations could indicate that there is a potential danger of anaphylactic reactions if anti-idiotypic antibodies are administered. In practice, however, this has not been observed in studies on mice and guinea pigs, the latter being particularly prone to developing anaphylaxis. Qualitative differences between heterologous and isologous anti-idiotypic antibodies might be the explanation.

Idiotopes carried by specific anti-allergen antibodies might be shared by antibodies of other specificities. This is part of the original concept of the idiotypic network (66) and has been confirmed in a number of systems, such as antibodies to autoantigens, like DNA (70). In animal models, antibodies to a defined idiotope lead to suppression of every antibody carrying this idiotope (63), irrespective of antibody specificity (54). However, when anti-idiotypic antibodies to an idiotype (a collection of idiotopes) are used, the induced suppression is apparently restricted to antibodies carrying the same set of idiotopes as the immunizing molecule used to produce the anti-idiotypic antibodies (86). Preferential suppression of antibodies carrying a set of idiotopes has been shown by using several monoclonal anti-idiotypic antibodies to idiotopes of that set (62). Moreover, suppression of the production of an antibody to an antigen does not seem to affect the production of antibodies with other specificities but of the same isotype: antibodies to a carrier protein are not suppressed as a result of idiotypic suppression of antibodies to a hapten (7). Nevertheless, in our studies of human immediate hypersensitivity, we have elected to use highly purified antiallergen antibodies to keep the suppression as specific as possible for the allergen.

CHARACTERISTICS OF THE IgE IMMUNE RESPONSE

The IgE immune response presents several characteristics which distinguish it from that of other isotypes. Since these have been recently reviewed (115), this discussion will focus on the reasons why the production of IgE antibodies might be more amenable to idiotypic suppression.

The total amount of IgE antibodies is small compared with other isotypes, even though a major proportion of IgE antibodies reside in tissues where they are bound to specific high-affinity receptors on mastocytes. The small amount of IgE is due to the very short half-life of soluble IgE antibodies (2.5 days) and to the small number of B lymphocytes committed to IgE synthesis. By limiting-dilution analysis it has been calculated that the frequency of these cells does not exceed 1 in 50,000 circulating B cells (103).

The production of IgE antibodies is strictly dependent on the presence of T cells. The formation of anti-hapten IgE in irradiated syngeneic recipients which had been transferred with primed cells depends on the presence of the same carrier as the one used to prime donors (56, 57). This opens possibilities of the regulation of IgE antibodies at different levels. For instance, anti-idiotypic antibodies specific for anti-carrier antibodies are particularly efficient in suppressing an anti-hapten IgE antibody production (8).

In addition, the regulation of IgE antibody production is not only under the control of *Ir* genes but also dependent on isotype-specific mechanisms in which suppressor T cells play a key role (30). This has been demonstrated in the conversion of low-IgE-responder strains of mice into high responders by low-dose irradiation or treatment with cyclophosphamide to eliminate suppressor T cells (25). In addition, the induction of suppressor T cells by denatured antigen has successfully been applied in the control of the IgE immune response (109). Since it is known that anti-idiotypic antibodies are potent inducers of suppressor T cells (34), there is at least strong circumstantial evidence that the use of anti-idiotypic antibodies could be potentially effective in the control of IgE responses as compared with other isotypes.

Furthermore, there is evidence in the literature suggesting a restriction in the number of possible idiotypes carried by IgE antibodies. The location of the C_ε heavy-chain gene toward the 5' end of DNA may limit the number of possible associations with V_H genes and D and J_H segments. Experimental support of this assumption has been obtained by analyzing the idiotype expression of anti-phosphorylcholine (PC) IgE antibodies during immunization (116). Attempts to maintain an idiotypic suppression of specific IgG antibodies to allergens usually fail after a few weeks; the emergence of antibody molecules carrying new idiotypic determinants is deemed to be responsible (110). In contrast, when IgE suppression is obtained, it can be maintained for long periods. Although this difference between IgE and IgG antibodies could be due to the relatively higher proportion of anti-idiotypic antibodies to IgE than to IgG, it could also be explained by a restriction of the number of possible V_H genes which can associate with the C_ε gene. The recent demonstration of a subset of B lymphocytes precommitted to the synthesis of both IgM and IgE antibodies carrying the same idiotype in the peripheral blood of nonatopic patients (75) further supports this view. The classical heavy-chain class-switching phenomenon through deletion during DNA rearrangement may not be operative for C_ε; expression of C_μ and C_ε may depend on an RNA-splicing mechanism (121). Idiotypes carried by IgE molecules may therefore express less diversity than IgG antibodies, a finding which has also been verified for IgM (40).

Idiotype diversity also arises from somatic mutations, which are very frequent on IgG antibodies during antigen-driven B-cell clonal expansion (76). Few data on the frequency of somatic mutations on IgE are available, but at least in experiments in which the IgE immune response toward a hapten such as *p*-azophenylarsonate (arsonate) has been analyzed, it was shown that the frequency of somatic mutations on IgE antibodies was lower than with other isotypes (55). Thus, compared with the other isotypes, mechanisms by which IgE antibodies could escape from idiotypic regulation might be less efficient.

ANIMAL MODELS OF IMMEDIATE HYPERSENSITIVITY

Models

The production of IgE antibodies by animals is rarely spontaneous. Experimental work on idiotypic suppression of IgE production has therefore invariably been carried out on induced IgE responses, in contrast to the spontaneous IgE production seen in human immediate hypersensitivity. The importance of this difference is discussed below.

Animal models which have been used for idiotypic suppression of an IgE immune response include mice and guinea pigs. Interspecies differences are further complicated by the broad nature of the immunizing material, which ranges from hapten-carrier conjugates, proteins, and peptides to specific or anti-idiotypic monoclonal or polyclonal antibodies of both isologous and heterologous origin.

I shall first describe active and passive immunization and then examine experimental data in the light of their impact on either a primary or an ongoing IgE immune response.

Experimental Designs

In vivo, suppression of idiotype expression can be achieved by two means. An animal can actively produce anti-idiotypic antibodies as a result of immunization with idiotype-carrying antibodies or cells (and in some cases with antigen) or, alternatively, receive soluble or cell-bound anti-idiotypic antibodies by passive transfer.

Active production of anti-idiotypic antibodies

Idiotypes are immunogenic; high titers of heterologous anti-idiotypic antibodies can be produced easily. However, depending on the animal species, or even the strain, used for anti-idiotype production, quantitative and qualitative differences in anti-idiotypic antibodies may appear. An immune response to a foreign idiotype is obviously comparable to a response to an antigen. This has been elegantly demonstrated in a recent study comparing heterologous and isologous anti-idiotypic antibodies in terms of diversity of the V_H, D, and J_H DNA segments used; whereas isologous anti-idiotypic antibodies are derived primarily from germline genes, heterologous anti-idiotypic antibodies are formed by selection of very diverse V_H segments (83). Qualitative differences between isologous and heterologous anti-idiotypic antibodies are important in the analysis of idiotopes (for example, in studies of idiotypic cross-reactivity) and for defining the physiological consequences of their use.

Although there was some early controversy, it now seems clear that the production of isologous anti-idiotypic antibodies is quite possible. Mice can probably produce anti-idiotypic antibodies to any isologous myeloma protein (101), and rabbits produce high titers of anti-idiotypic antibodies as a result of inoculation with antibodies from allotype-matched animals (119). Spontaneous production of anti-idiotypic antibodies during a normal immune response has been repeatedly observed (93). It has even been suggested that idiotypes carried by isologous antibodies are more immunogenic per se than idiotypes carried by heterologous antibodies (13). Allotypic mismatching is not required for efficient anti-idiotypic antibody induction (101).

Idiotype immunogenicity can be increased by conventional means such as the use of Freund adjuvant (5). One alternative to the use of classical adjuvants is to form complexes by mixing the antibody with the corresponding antigen; complexes in antibody excess are efficient inducers of anti-idiotypic antibodies (69). Idiotype immunogenicity may also be increased through polymerization or clonal expansion of B lymphocytes carrying the relevant specific antibody before injection. However, the nature of induced anti-idiotypic antibodies differs with the immunization protocol. Soluble antibodies in free form induce preferentially the production of anti-idiotypic antibodies to idiotopes included in the paratope, whereas antibodies in the form of complexes with antigen induce a preferential production of anti-idiotypic antibodies to idiotopes of the framework (69). In addition, the binding of an antigen to an antibody generates new idiotypic determinants, which can be identified by monoclonal antibodies (88). The situation is further complicated, however, by variations in the nature of anti-idiotypic antibodies produced depending on the animal model used. For instance, guinea pigs immunized with free isologous anti-BPO antibodies produce a majority of antibodies to framework idiotopes (118), whereas mice immunized with free isologous anti-BPO-specific antibodies produce mainly anti-idiotypic antibodies to paratope-associated idiotopes (6).

Efficient in vivo production of anti-idiotypic antibodies has been achieved by inoculation with purified antibodies of both monoclonal and polyclonal origin (5, 14, 45). The latter can either be used in soluble form or as surface immunoglobulin on B lymphocytes (42) or covalently coupled to cells (63). Specific antibodies to proteins (78), haptens (5), carriers (8), or synthetic polymers (31) have also been successfully used.

Finally, inoculation of antigen itself induces the production of anti-idiotypic antibodies (41), particularly when it is administered in polymerized form or covalently coupled to cells (10). In addition, when IgE antibody production has been suppressed, new exposure to antigen may increase the production of anti-idiotypic antibodies (5), thus maintaining the state of suppression. Although this has not been studied in any detail in experimental IgE-mediated immune responses, part of the IgE suppression induced by inoculation of high-molecular-weight antigen derivatives (120) might be related to the induction of anti-idiotypic antibodies.

Analysis of the kinetics of anti-idiotypic antibody production has demonstrated that anti-idiotypic antibodies may precede the appearance of specific antibodies to the initial antigen. In the B-cell proliferative response to trypanosomiasis, anti-idiotypic antibodies were detected as early as 3 days after infection, whereas primary antibodies appeared only after day 6 (93).

Moreover, an inverse relationship has been observed on a number of occasions between the level of specific antibodies and corresponding anti-idiotypic antibodies in serum. Thus, the level of anti-idiotypic antibodies to anti-alprenolol antibodies varies in inverse proportion to that of the anti-alprenolol antibodies (28). Anti-idiotypic antibodies precede the appearance of anti-PC antibodies in the course of immunization (23). Cyclical variations in anti-idiotypic and specific antibody concentrations suggest that antigen disrupts a preestablished equilibrium between antibody pairs. However, analysis of the antibody response to bacterial levan in athymic mice, which are unable to mount an anti-idiotypic response, has challenged the concept of a direct influence of anti-idiotypic antibodies on the production of specific antibodies (61).

As already discussed (3), the use of an antigen to induce anti-idiotypic antibodies has a major advantage over other methods, inasmuch as it can suppress the production of the entire immune response to an antigen; i.e., it can suppress the expression of all the idiotypes carried by specific antibodies. This advantage may turn out to be of practical relevance, since immunodominant or regulatory idiotopes have not been identified, except on a few occasions (54). However, experimental evidence indicates that an immune response to an antigen can be almost completely suppressed by using a few representative specific antibodies (60). It should be pointed out, however, that in this case the precise mechanism by which production of other antibodies to the same antigen is suppressed remains unclear and may be unrelated to idiotypic suppression.

Passive transfer of anti-idiotypic antibodies

A second method of suppressing the expression of an idiotype is the passive administration of anti-idiotypic antibodies. Sera containing anti-idiotypic antibodies or affinity-prepared anti-idiotypic antibodies have been transferred (9, 45), either in soluble form or covalently coupled to cells. Anti-idiotypic antibodies of both isologous (118) and heterologous (63) origin have been used.

The administration of anti-idiotypic antibodies can result in different effects on the immune response, depending on the amount of antibodies given (68), presumably on the isotype of anti-idiotypic antibodies (92), and on their complement-activating capacities (33). The route of injection may also influence the effect on the immune response; anti-idiotypic antibodies given by the subcutaneous route may enhance the expression of the corresponding idiotype, whereas intravenous administration usually leads to suppression (3). Moreover, it is not clear whether anti-idiotypic antibodies to framework idiotopes, as opposed to antibodies to paratope-associated idiotopes, are preferentially involved in regulation, which might explain the unexpected divergences between results obtained in different experimental models. On a practical level, the fact that the transfer of relatively high doses of anti-idiotypic antibodies has to be repeated frequently to maintain idiotype suppression would appear to limit potential therapeutic applications.

An alternative to the direct transfer of anti-idiotypic antibodies has been developed by Malley et al. Spleen cells from virgin mice are incubated in vitro with heterologous rabbit anti-idiotypic antibodies (78). This procedure induces the expansion of specific suppressor T cells, which can then be passively transferred to recipients (80).

Suppression of IgE Antibody Production

There is a clear distinction between the ease with which a primary IgE immune response can be suppressed idiotypically and the more difficult situation encountered with established IgE responses. Since this distinction is crucial for human immediate hypersensitivity, data on the prevention of IgE production and attempts to control an ongoing IgE immune response are reviewed separately.

Suppression of a primary IgE immune response

Active production of anti-idiotypic antibodies. *Mouse model.* The idiotypic regulation of a primary mouse IgE immune response has been studied mainly by de Weck's group and was partly summarized a few years ago (3). BALB/c mice were inoculated by the subcutaneous or intraperitoneal route with purified isologous anti-BPO antibodies in Freund adjuvant. Active production of specific anti-idiotypic antibodies was detected after week 6 postinoculation. In these mice, immunization with BPO-ovalbumin (OVA) together with $Al(OH)_3$ failed to induce IgE production, even on repeated inoculations of BPO-OVA (5), whereas control mice produced high titers of anti-BPO IgE as evaluated by PCA in the rat skin. Suppression of IgE production was specific for the hapten, since the response to OVA was not affected (7). Interestingly, inoculation of antigen boosted the production of anti-idiotypic antibodies. These results have been partly confirmed in studies of the IgE immune response to PC (2), in that a partial reduction of IgE production was obtained in mice actively producing antibodies to T15, a major idiotype of the anti-PC immune response. The interpretation of these experiments is however complicated by the fact that the IgE response of control mice could not be maintained at its highest level, despite repeated inoculations with PC-keyhole limpet hemocyanin in $Al(OH)_3$. Another difficulty arose from the subsequent demonstration that IgE antibodies belonged to group II anti-PC antibodies, which do not carry the T15 idiotype (116); group II anti-PC antibodies utilize V_H genes distinct from group I antibodies (24).

Although these results indicate that specific anti-BPO or anti-PC IgG antibodies share idiotypic determinants with IgE of the same specificity, the pattern of specific IgG antibody production was variable in mice forming anti-idiotypic antibodies. In the BPO system, a moderate rise in specific IgG levels was observed early, whereas in control mice the rise in specific IgG occurred only after week 6 but reached concentrations three to six times higher than those in the test mice (5, 6). In the PC system (4), reduction of specific IgG production varied according to the subtype, whereas IgA antibodies were not affected by anti-T15 antibodies, although they carried the T15 idiotype. It is likely that in this case, other minor idiotopes expressed on the T15$^+$ antibodies used to induce anti-idiotype production are involved in the suppression mechanism; these minor idiotopes may be absent on IgA but carried by IgE. For another system, the immune response to arsonate, it was shown that anti-idiotypic antibodies to minor idiotopes were able to suppress idiotype expression as efficiently as anti-idiotypic antibodies toward major idiotopes (51). It is worth remembering that anti-idiotypic antibodies produced by BALB/c mice are mainly, but not exclusively, directed to paratope-associated idiotopes, whereas in the guinea pig model with the same hapten, most anti-idiotypic antibodies are directed to framework idiotopes.

Immunization with B cells carrying the relevant idiotype or with idiotype-coupled cells is an effective method of induction of anti-idiotypic antibodies. However, different results are observed, depending on the route used for immunization. Intravenous (3) or intraperitoneal (63) inoculation leads to suppression of idiotype expression, whereas subcutaneous administration of cells regularly enhances the production of antibodies carrying the corresponding idiotype (3).

De Weck's group have extended their investigations to the anticarrier immune response. They reasoned that since IgE production is strictly dependent on carrier-specific T cells, production of anti-idiotypic antibodies to anticarrier antibodies might result in the suppression of IgE to both carrier and hapten, provided that the latter was coupled to the same carrier for subsequent immunization. Indeed, BALB/c mice actively producing anti-idiotypic antibodies to anti-OVA antibodies are unable to mount an IgE immune response to the BPO-hapten and, to a lesser extent, to OVA (8). Similar results were obtained with OVA substituted with 2,4-dinitrophenyl, but the IgE immune response to BPO was not reduced if *Ascaris* proteins instead of OVA were used as a carrier. Specific anti-BPO suppression with anticarrier anti-idiotypic antibodies lasted longer than when anti-idiotypic antibodies to anti-hapten antibodies were used.

Guinea pig model. Initial experiments with guinea pigs had shown that alloantisera raised against lymphoid cells were able to inhibit in vitro the antigen-specific proliferation of T cells to a variety of antigens (42). This suggested that alloantibodies recognized specific receptors on T cells, since the serum inhibitory activity was adsorbed by lymphoid cells bearing recognition structures for the corresponding antigen (43). Later, isologous anti-idiotypic antibodies were prepared against purified anti-BPO IgG antibodies; these anti-idiotypic antibodies were able to inhibit in vitro the antigen-specific T-cell proliferation in a strictly isologous manner, indicating that antigen-specific T cells carried strain-specific recognition structures similar to antibody idiotype (44).

Guinea pigs actively producing anti-idiotypic antibodies after immunization with syngeneic antibodies against BPO coupled to bovine gamma globulins (BGG) are unable to mount a normal reaginic antibody response to BPO upon subsequent immunization with BPO-BGG in Al(OH)$_3$ (45). IgE levels, measured by PCA, were still depressed 9 weeks after immunization, even though repeated booster immunizations were given. Compared with the mouse model, the suppressive state seemed to be more easily maintained (118). Unfortunately, no data on the evolution of specific IgG antibodies were made available for this model. In contrast to the mouse, inoculation of cells carrying specific antibodies has been more uniformly associated with a prolonged state of suppression of the primary IgE immune response (3).

Passive transfer of anti-idiotypic antibodies. *Mouse model.* Passive transfer of anti-idiotypic antibodies has been performed with isologous and with heterologous antibodies. Mice which are primed with *Ascaris* proteins in Al(OH)$_3$ and which have received isologous anti-idiotypic antibodies to anti-BPO antibodies produce anti-BPO IgE antibodies upon immunization with BPO-*Ascaris* proteins only after a delay corresponding to the decrease of anti-idiotypic antibodies in the serum (5). Intravenous administration of anti-T15 isologous antiserum completely prevented the formation of IgE antibody to PC-keyhole limpet hemocyanin. Three injections of anti-T15 antiserum resulted in suppression of IgE formation which lasted for at least 150 days (6). Suppression was specific for T15$^+$ antibodies, since anti-idiotypic antibodies raised to T15$^-$ anti-PC myeloma proteins had no effect (2). In these experiments, controls were used to ensure that anti-idiotypic antibodies did not inhibit the PCA reaction used to evaluate IgE antibodies.

An efficient control of IgE production has also been obtained by the transfer of anti-idiotypic antibodies specific to anticarrier antibodies. A single injection of 0.2 ml of anti-idiotypic antiserum to anti-OVA antibodies suppressed IgE formation for at least 12 weeks despite additional antigen inoculation, provided that the same carrier was used for boosting (9). The amount of anti-idiotypic antibodies which was used for each injection corresponded to 1 to 3 µg of antibodies, of which up to 70% recognized the paratope of the specific antihapten antibodies (5).

The effect of passive transfer of isologous antibodies on IgG production in the anticarrier suppression system has been examined (9). A significant reduction in IgG production was seen only after 6 weeks, compared with control mice receiving normal serum instead of anti-idiotypic antibodies; this delay may be related to the longer half-life of IgG (21 days) compared with IgE (2.5 days).

Heterologous anti-idiotypic antibodies, usually from rabbits or guinea pigs, have been used in the IgE immune response to hapten or proteins. A/J mice produce a family of idiotype-related antibodies in response to immunization with arsonate, including a cross-reactive idiotype on specific IgE antibodies (64). Transfer of rabbit anti-idiotypic antibodies raised against monoclonal antiarsonate antibodies resulted in complete suppression of the antiarsonate immune response, including IgE (63). Specific IgE antibodies raised in mice to antigen B (AgB) of Timothy grass pollen have been used to raise anti-idiotypic antibodies in rabbits (78). These anti-idiotypic antibodies, passively transferred to recipients, induced a complete lack of anti-AgB-specific IgE formation for at least 35 days, despite antigen inoculation boosts (79). It was further shown by the same investigators that in vitro incubation of virgin mouse spleen cells with rabbit anti-idiotypic antibodies and subsequent transfer of spleen cells to naïve syngeneic recipients was sufficient to induce a state of IgE unresponsiveness (80).

Guinea pig model. In the guinea pig model, transfer of isologous anti-idiotypic antibodies on four consecutive days following a second inoculation of BPO-BGG in Al(OH)$_3$ suppressed the production of IgE antibodies for at least 6 weeks, even though a booster inoculation was given (118). A total of 4 mg of an immunoglobulin fraction containing anti-idiotypic antibodies had to be given for each animal. Anti-BPO IgG and IgM antibodies were also suppressed temporarily by intravenous or subcutaneous administration of anti-idiotypic antibodies (45).

Suppression of an ongoing IgE immune response

The evidence for idiotypic regulation of an ongoing IgE immune response is far less convincing, perhaps because long-term IgE production in animals is not easy to establish. Passive administration of isologous anti-idiotypic antibodies in mice sensitized with BPO-OVA results in a short-term depression of IgE levels. Repeated inoculations of anti-idiotypic antibodies can extend the suppressive state, which remains limited to the persistence of anti-idiotypic antibodies in the serum. However, a long-standing suppression of IgE production has been achieved for anti-PC IgE with anti-T15 anti-idiotypic antibodies (6).

Inoculation of anti-idiotypic antibodies to anticarrier antibodies has been more efficient in the long-term suppression of an ongoing IgE production. Mice producing high titers of IgE antibodies to BPO were given intravenous inoculations of 0.2 ml of serum containing 20 to 60 µg of anti-idiotypic antibodies to anticarrier (OVA) antibodies. Suppression of IgE production was maintained for several weeks, despite reinoculation of BPO-OVA (9).

Attempts to suppress an ongoing IgE immune response in guinea pigs immunized to the BPO-hapten have included the use of both idiotype-bearing and anti-idiotypic antibodies. Guinea pigs actively producing IgE antibodies and inoculated with autologous immune serum (four 0.5-ml intradermal administrations) with or without Freund complete adjuvant showed no change in IgE production, except for the nonspecific effect of the adjuvant itself (45). Unfortunately, results of inoculation of purified autologous anti-BPO-BGG antibodies were inconclusive, since the PCA titer was difficult to maintain in these animals. Addition of various quantities of BPO-BGG to immune serum to form immune complexes before intradermal inoculation usually resulted in a boost of IgE production. However, the presence of xenogeneic immunoglobulins in the immunizing material makes these results difficult to interpret, since immune complexes containing heterologous immunoglobulins enhance the immune response toward the antigen (111).

Repeated administration of anti-idiotypic antibodies induced a progressive decline in anti-BPO IgE antibody production, which could be observed 3 weeks after anti-idiotype inoculation. However, a subsequent boost injection with antigen overcame the suppressive state (118). Moreover, the anti-BPO IgG response was unaffected or only marginally affected.

Transfer of anti-idiotypic antibodies of heterologous origin is particularly efficient in suppressing an established IgE immune response. Rabbit anti-idiotypic antibodies raised to mouse IgE antibodies specific for antigen B (78) were able to suppress an established IgE response (79). Transfer of in vitro anti-idiotypic-stimulated spleen cells also suppressed IgE production (80).

Mechanism of Action of IgE Immune-Response Suppression

A direct interaction between anti-idiotypic antibodies and B cells carrying the corresponding idiotype has been described in several experiments. Anti-idiotypic antibodies can suppress idiotype expression by elimination of idiotype-bearing cells (33). This effect depends on the use of an appropriate quantity of anti-idiotypic antibodies and requires Fcγ fragment-dependent functions, namely, complement activation and macrophage-fixing capacities. Evidence for suppression at the B-cell level has also been presented for isologous anti-idiotypic antibodies (92). Anti-idiotypic antibodies to framework idiotopes may be more efficient in this respect (68). However, other experiments have shown that direct and highly selective priming of B cells can be produced by anti-idiotypic antibodies (36). In vitro, rabbit anti-idiotypic antibodies enhance, in a dose-dependent manner, the production of antibodies carrying the corresponding idiotype (112). Transfer of anti-idiotypic antibodies to sensitized animals results in a functional inhibition of IgE-producing B cells which, if the anti-idiotypic antibody inoculations are repeated, may lead to cell deletion (3).

T cells of the helper/inducer phenotype carry surface receptors which resemble idiotypes sufficiently to interact directly with anti-idiotypic antibodies. Specific priming of helper T cells by anti-idiotypic antibodies has been shown both in vitro (117) and by transfer experiments followed by the generation of a secondary immune response to group A streptococcal carbohydrate in the recipient (36). Data on the suppression of IgE production with anti-idiotypic antibodies to carrier-specific antibodies indicate that anti-idiotypic antibodies interact directly with helper T cells, producing at least a functional impairment of T-B cell cooperation.

Finally, induction of antigen-specific suppressor T cells by anti-idiotypic antibodies has been demonstrated by cell transfer experiments in the anti-hapten IgE response (3) and in the suppression of IgE production by heterologous anti-idiotypic antibodies both in vitro (80) and in vivo (63, 79). Induction of suppressor T cells is Fc dependent and requires the presence of macrophages presenting anti-idiotypic antibodies to histocompatible T cells (77). However, a segment of the third hypervariable loop of the V_L domain of a BALB/c myeloma protein has been shown to stimulate anti-idiotypic T suppressor cells (102). A dose-dependent effect can also be observed; low doses of anti-idiotypic antibodies may induce a long-lasting state of suppression, whereas higher doses produce an immediate but transient suppression (34).

The higher susceptibility of a primary IgE immune response to idiotypic suppression could be due merely to the smaller number of precommitted B cells or to a higher sensitivity of precursor cells (36). The emergence of new idiotypic determinants during the antigen-driven clonal expansion could also explain why IgE antibodies produced during an ongoing immune response might escape suppression. This seems to be a likely explanation for the lesser impact of anti-idiotypic antibodies on IgG production, but there is little evidence for idiotypic shift of IgE antibodies (see above).

Summary

IgE antibody production can be suppressed in mice and guinea pigs either by active production of anti-idiotypic antibodies or by passive transfer of these antibodies, be they of isologous or heterologous origin.

Anti-idiotypic antibodies can be produced through repeated conventional immunization with specific antibodies. Animals producing anti-idio-typic antibodies are unable to mount a specific IgE immune response upon subsequent immunization with alum-adsorbed proteins or hapten-carrier conjugates. Transfer of anti-idiotypic antibodies of isologous origin results in only a short-term prevention of IgE production unless anti-idiotypic antibodies are given repeatedly and at a sufficiently high dose.

However, the suppression of an ongoing IgE response is much more difficult to achieve. Repeated transfer of isologous anti-idiotypic antibodies or use of heterologous anti-idiotypic antibodies has usually been successful. The production of anti-idiotypic antibodies to anticarrier antibodies or the transfer of these antibodies to recipients has also been attempted, with some positive results. The state of suppression, however, remains transient in most cases, since a subsequent boost with antigen usually restores IgE production.

The modulation of idiotypic expression by isologous anti-idiotypic antibodies has obvious importance in the study of the mechanisms regulating specific IgE production. Moreover, autologous idiotype–anti-idiotype interactions are clearly of greater interest than heterologous systems in terms of ultimate therapeutic applications of IgE suppression in humans.

HUMAN IMMEDIATE HYPERSENSITIVITY

From Animal Models to Human Immediate Hypersensitivity

IgE production in animals is influenced by genetic factors. Low- and high-responder mouse strains have been identified (114). A close association has been shown between the capacity of humans to mount an IgE immune response to certain pollen allergens and the presence of class II major histocompatibility complex antigens (81). The absence of correlation between symptoms of immediate hypersensitivity and the level of anti-grass pollen antibodies adds further support to a relationship between IgE responsiveness and the genetic background (107). In addition, in both animals and humans, a peripheral control of IgE production is exerted by suppressor T cells whose function may be inhibited in animal studies by cyclophosphamide treatment or low-dose irradiation (56).

Despite these similarities between animal models of IgE production and human immediate hypersensitivity, special care should be taken before extrapolating experimental results to humans. An IgE immune response is induced in animals through intraperitoneal or subcutaneous inoculation, whereas in humans, with the exception of hypersensitivity to Hymenoptera insect venom and possibly atopic dermatitis, sensitization takes place through the mucosa of the respiratory or gastrointestinal tract, with IgE antibodies being produced in lymphoid tissues associated with these mucosa (108); the main sites of IgE production in animals are the spleen and lymph nodes. The mucosal immune system can be considered a partly independent entity, whose organization differs from that of the systemic immune system. Few data have been gathered on antibody regulation in the mucosa, but those that do exist suggest that different regulation mechanisms of antibody production might operate (105). Conventional hyposensitization is efficient in systemic hypersensitization, e.g., Hymenoptera insect venom (17), but its effect on the bronchial immune system is less clearly established.

No matter what material is used to induce IgE production in animals, be it an antigen, a hapten-carrier conjugate, or a polypeptide, the antibody production is highly dependent on the adjuvant used. Classical adjuvants such as Freund adjuvant will preferentially lead to production of IgG antibodies, whereas IgE is produced if the immunizing material is adsorbed on aluminum hydroxide or given together with *Bacillus subtilis* or if the animal is infested with parasites such as *Nippostrongylus brasiliensis* (113). In humans, IgE production develops in the absence of any adjuvant, although it is well known that parasitic infestations or viral infections may enhance IgE responsiveness (89).

The animal IgE immune response is, in most cases, transient and has to be maintained by regular inoculations of the immunogen. In humans, a single exposure of a sensitized individual to an allergen may trigger IgE production for very long periods, without the need for reexposure. Such differences lead one to wonder whether an ongoing IgE immune response induced in animals can be compared with the long-term steady-state IgE production in humans.

Most IgE immune responses in man are directed toward a number of allergenic determinants, and, hence, one may expect a great diversity of idiotypes carried by IgE antibodies, with consequent difficulty in suppressing the immune response. In practice, however, this may well not be the case. Thus, in the human immune response to *D. pteronyssinus* allergens, we have found that idiotypes expressed on IgE antibodies of unrelated individuals cross-reacted to a large extent, suggesting either that these idiotypes were derived from germ line genes or that they resulted from antigen-driven somatic mutations (99). The recent demonstration of a subset of B cells committed to the synthesis of both IgE and IgM (75), indicating that classical isotypic switching is not involved in IgE production, adds some support to the first of these hypotheses. Expression of different idiotypes on IgG and IgM antibodies to streptococcal A carbohydrate was already considered a few years ago as an indication that different pre-B cells were committed to either IgG or IgM production (37).

In addition, although multiple hypersensitivities are usually demonstrated by the assay of circulating IgE antibodies, the clinical relevance of those remains questionable. In a majority of cases, there are only a few allergens of principal clinical interest; this reduces the diversity of the immune response to be suppressed. Even if it turns out that the IgE immune response in humans is characterized by a large diversity of idiotypes, this does not necessarily mean that idiotypic suppression cannot be achieved. Indeed, the experimental evidence reviewed above indicates that a polyclonal immune response to an antigen can be suppressed, provided all the relevant antibodies are used for anti-idiotypic antibody induction. In addition, it is possible that suppression of antibody production to a single epitope results in the suppression of antibodies to all epitopes of the same molecule (60).

A practical difficulty in exploring the feasibility of idiotypic suppression of IgE antibody production in humans arises from the very low concentration of IgE antibodies in serum; this effectively precludes its preparation in sufficient quantity and purity for use in the induction of anti-idiotypic antibodies. The way in which this difficulty has been overcome is described in Active Production of Anti-Idiotypic Antibodies in Human Immediate Hypersensitivity.

Human Anti-Idiotypic Antibodies

Before examining the means by which IgE production might be regulated by idiotype–anti-idiotype interactions, I will review the evidence indicating that regulation at the idiotype level may exist under physiological conditions. Three questions should be addressed. First, are anti-idiotypic antibodies produced in immediate-hypersensitivity reactions? Second, what are the reasons to believe that anti-idiotypic antibodies play a role in IgE regulation? Finally, what is the situation of nonatopic individuals with respect to anti-idiotypic antibodies, given that they do not produce sufficient IgE to develop hypersensitivity symptoms?

Are auto-anti-idiotypic antibodies produced in immediate hypersensitivity?

Several studies on allergic individuals have documented the presence of anti-idiotypic antibodies, as a result of either spontaneous production of or a booster immunization with allergen. Anti-idiotypic antibodies have been identified in the serum of one individual who was allergic to *Lol pI* grass pollen and had been treated by conventional hyposensitization (16). These antibodies were partially purified and were shown to inhibit in a dose-dependent and antigen-specific manner the binding of specific anti-*Lol pI* antibodies to allergens. Antibodies able to bind specific anti-ragweed antibodies have also been detected in the sera of atopic patients (20). Binding of these anti-idiotypic antibodies to purified anti-ragweed antibodies could be inhibited by ragweed but not by unrelated allergens (21). However, absolute proof of the anti-idiotypic nature of these antibodies cannot be established, since removal of the inhibiting activity by adsorption on allergen-specific antibodies was not carried out.

In our own studies on hypersensitivity to *D. pteronyssinus,* we have found antibodies which fulfilled criteria for anti-idiotypic antibodies: (i) they bound specifically to purified anti-*D. pteronyssinus* antibodies of both IgG and IgE isotypes, evaluated in separate immunoassays; (ii) this binding was inhibited by up to 80% by the addition of free *D. pteronyssinus* allergens, whereas addition of unrelated allergens had no effect; (iii) adsorption on F(ab')$_2$ fragments of anti-*D. pteronyssinus* IgG completely removed their binding to IgG, but only partially removed their binding to anti-*D. pteronyssinus* IgE; and (iv) antibodies whose binding to specific IgE could be inhibited by addition of allergen could not be adsorbed on insolubilized specific IgG antibodies. These findings indicated that at least for the 10 atopic individuals investigated, who were not under hyposensitization treatment, the serum contained antibodies able to bind specific anti-*D. pteronyssinus* IgG and IgE antibodies at the level of framework and para-tope-associated idiotopes (97): however, anti-idiotypic antibodies directed to paratope-associated idiotopes were different for IgE and IgG antibodies. Further purification of the anti-idiotypic antibodies confirmed that a majority were directed toward the paratopes of specific anti-*D. pteronyssinus* antibodies and that they inhibited in a dose-dependent manner the binding of these antibodies to *D. pteronyssinus* (98).

The question of the influence of hyposensitization on the level of anti-idiotypic antibodies has been investigated in studies on ragweed-hypersensitive patients. Immunoglobulins prepared from the sera of hyposensitized patients showed a higher degree of binding to a monoclonal anti-ragweed antigen E antibody which cross-reacts with human anti-antigen E antibodies than did immunoglobulins prepared from nonhyposensitized patients (22). Unfortunately, in these assays no attempt was made to separate specific antibodies from anti-idiotypic antibodies in the serum. The level of antibod-ies able to bind to radiolabeled human anti-*Lol pI* F(ab')$_2$ fragments was monitored during pollen exposure and in the course of specific immuno-therapy (15). Although a general pattern of anti-idiotypic antibody evolu-tion did not emerge from these studies, it appeared that the level of anti-idiotypic antibodies rose following seasonal exposure and immunotherapy. It has also been our experience that evolution of anti-idiotypic antibodies in the human immune response is difficult to evaluate unless both purified anti-allergen antibodies and anti-idiotypic antibody containing fractions depleted of anti-allergen antibodies are used. Booster immunization with other antigens, such as TT, has also resulted in an increase of anti-idiotypic antibodies (47) and in the appearance of B cells bearing anti-idiotypic surface immunoglobulins (48).

Do anti-idiotypic antibodies play a role in human immediate hypersensitivity?

A large body of evidence suggests that anti-idiotypic antibodies play an important role in non-IgE immune responses. Sheep anti-idiotypic antibodies raised against a monoclonal human IgM antibody were able to suppress specifically the in vitro production of IgM by B cells carrying the same surface immunoglobulin (12). Rabbit anti-idiotypic antibodies to a monoclonal human IgG isolated from the serum of a child with hypogammaglobulinemia inhibited the spontaneous in vitro production of the antibody (85). A cross-linked monoclonal antibody to an idiotype expressed on specific anti-N-acetylglucosamine antibodies induced in vitro the production of specific IgM by purified B cells (11). Rabbit anti-idiotypic antibodies carrying an internal image of the Fcγ fragment induced human peripheral blood lymphocytes to proliferate and suppressed the pokeweed mitogen-dependent production of rheumatoid factor (39).

With regard to IgE immune responses, it has been shown that rabbit anti-idiotypic antibodies prepared against monoclonal anti-*Lol pI* antibodies and recognizing idiotypic determinants of human specific antibodies (84) induce the release of histamine when added to IgE-coated basophils isolated from the peripheral blood of patients hypersensitive to ryegrass (82). More directly, a regulatory role for anti-idiotypic antibodies in the production of IgE antibodies has been suggested by results of experiments in which peripheral-blood B lymphocytes from three TT-boosted patients produced IgE. Addition to the cell culture of rabbit anti-idiotypic antibodies raised to anti-TT IgG antibodies suppressed IgE production in a dose-dependent manner (49), indicating that IgG and IgE anti-TT antibodies cross-reacted at the level of their idiotypes (50). This suppression was more efficient when induced in an autologous way. Specific anti-TT IgG antibodies were also partly suppressed in such a system.

The first in vivo demonstration that anti-idiotypic antibodies might be involved in immediate-type hypersensitivity came from the observation that rabbit anti-idiotypic antibodies could elicit a Prausnitz-Küstner reaction in normal skin previously sensitized with IgE antibodies from an allergic donor (46). From such findings, it has been suggested that anti-idiotypic antibodies could be responsible for the maintenance of anaphylactic reactions after allergen exposure. However, this has not been confirmed, nor is it known whether isologous anti-idiotypic antibodies would behave in the same way as the heterologous antibodies described above. We have observed an inverse relationship between the concentrations of specific antiallergen antibodies and the corresponding anti-idiotypic antibodies in atopic individuals not undergoing hyposensitization. This has been verified for both specific IgG and IgE anti-*D. pteronyssinus* antibodies evaluated independently (M. G. Jacquemin and J. M. R. Saint-Remy, unpublished data). Bose et al. have shown the same type of relationship in the sera of nondesensitized allergic and hyposensitized patients (15). Moreover, in our clinical studies (see below), an increase in the level of anti-idiotypic antibodies was observed while the level of specific antiallergen antibodies decreased.

Taken together, the above studies suggest that anti-idiotypic antibodies do regulate the production of antiallergen antibodies, although no solid evidence for their involvement in perpetuating allergic reactions has been produced.

Do nonatopic individuals produce anti-idiotypic antibodies?

It has been proposed that nonatopic individuals do not produce high levels of IgE antibodies, because of an efficient regulation by anti-idiotypic antibodies (22). Indeed, conventional hyposensitization results in an increase in the level of anti-idiotypic antibodies to that prevailing in nonatopic subjects. Others (15) have failed to confirm this distinction between atopic and nonatopic individuals on the basis of a lower ratio of anti-idiotypic antibodies to specific antiallergen antibodies. We have found that all sera of nonatopic individuals assayed so far contained anti-*D. pteronyssinus* antibodies of every isotype, including IgE. The levels of specific IgG, IgA, and IgM antibodies were 2- to 3-fold lower than in atopic individuals, while the concentration of specific IgE antibodies was 100-fold lower (100). Anti-idiotypic antibodies to antiallergen antibodies were also found, but the ratio of specific antibodies to corresponding anti-idiotypic antibodies was the same as in atopic individuals. Moreover, like atopic individuals, nonatopic individuals produce a majority of anti-idiotypic antibodies to paratope-associated idiotopes; a dose-dependent inhibition of allergen binding to specific antibodies has been found (64a).

Idiotypic Modulation of Human Immediate Hypersensitivity

Anti-idiotypic antibodies constitute a normal component of the human immune response to allergens. Natural exposure to the latter or their inoculation into sensitized patients shifts both the specific antibodies and the corresponding anti-idiotypic antibodies toward higher levels. Such observations inevitably lead to speculation about the possibility of increasing the production of anti-idiotypic antibodies in the hope of reducing the production of antiallergen antibodies. How can this be achieved?

As mentioned previously, certain forms of antigen might preferentially enhance anti-idiotypic antibodies (120), but experience with these derivatives in humans has not been convincing until now. The therapeutic use of passive transfer of anti-idiotypic antibodies to allergic individuals might be envisaged. However, both the amount and the isotype of anti-idiotypic antibodies can influence the outcome of anti-idiotypic antibody administration; too small an amount may boost the anti-allergen immune response, including IgE antibodies. Relatively large amounts of anti-idiotypic antibodies would need to be transferred regularly, raising the practical problems of how to obtain these antibodies, since the amounts of anti-idiotypic antibodies which can be isolated from the sera of atopic donors are exceedingly small (98). Pools of gamma globulins could in theory be selected for their content of specific anti-idiotypic antibodies to antiallergen antibodies of a given specificity, in much the same way as hyperimmune sera are prepared for passive immunotherapy to infectious diseases. These pools of selected immunoglobulins might contain concentrations of anti-idiotypic antibodies sufficient to reduce IgE antibody levels. Combined passive and active immunotherapy for allergic diseases has been advocated as more efficient than active immunization alone (1). At least in one other clinical condition, i.e., acquired hemophilia, in which patients develop an autoimmune response to factor VIII, perfusion of selected pools of immunoglobulins containing anti-idiotypic antibodies to anti-factor VIII antibodies results in a suppression of anti-factor VIII antibody production (94).

Finally, a selective increase in anti-idiotypic antibody production could be achieved by active immunization with specific antiallergen antibodies. It would, however, appear to be necessary to increase the immunogenicity of idiotypes carried by these antibodies to elicit specific anti-idiotypic antibodies. In the following sections I will report our experience and first clinical results in this field.

Active Production of Anti-Idiotypic Antibodies in Human Immediate Hypersensitivity

Specific antiallergen antibodies

Precise evaluation of antiallergen antibodies in serum is difficult. Quantitative immunoassays for specific IgE antibodies are available (52), but interference due, for instance, to the competitive binding of IgG antibodies to insolubilized allergen (53) makes quantification difficult. With a few exceptions (59), assays for serum antiallergen IgG or IgA antibodies are even less easy to interpret, mainly because of nonspecific adsorption to the solid phase used for allergen insolubilization. To evaluate more precisely the quantities of specific antibodies available, we have prepared specific antibodies to *D. pteronyssinus* (100) or to grass pollen allergens by immunoadsorption. This method measures low-affinity antibodies which escape detection by conventional immunoassays and which may be important for the induction of anti-idiotypic antibodies. From the sera of 16 severely ill asthmatic patients who were sensitive to *D. pteronyssinus* and who had been previously desensitized, we extracted specific antibodies at an average concentration of 40 µg/ml of serum; the isotypic distribution of these antibodies was as follows: 52% IgG, 40% IgM, 8% IgA, and 0.1% IgE. In non-desensitized allergic individuals, the average concentration of specific antibodies was 10 µg/ml, with a similar isotypic distribution. For grass pollen-hypersensitive patients, the average concentration of specific antibodies recovered by adsorption was 6 µg/ml. Sufficient quantities of specific antibodies for both analytic and therapeutic purposes can thus be recovered by adsorption.

However, it is clear from the isotypic distribution that if one is limited to the use of specific IgE antibodies only, the amounts of antibodies would be insufficient, even if they could be isolated with acceptable purity. We therefore examined the extent to which the degree of idiotypic cross-reactivity between IgE and IgG antibodies was sufficient to allow the use of the more abundant specific IgG antibodies as substitutes for IgE. Previous studies had indicated that human IgG and IgE antibodies cross-reacted significantly in both the TT (50) and ryegrass pollen (16) systems, a finding which fits current concepts of isotypic switching. However, these studies did not distinguish idiotopes associated with the antigen-binding site of specific antibodies from those of the framework region of the antibody. We have carried out such analyses with rabbit (99) or autologous (98) anti-idiotypic antibodies and have shown that for a single individual, IgG and IgE antibodies to the same allergen share idiotopes of the framework but not idiotopes associated with the allergen-binding site. Therefore, it seemed possible to use specific IgG antibodies instead of specific IgE antibodies at least for the induction of an anti-idiotypic antibody response to framework idiotopes.

We have also considered the use of pools of immunoglobulins enriched in specific antiallergen antibodies of defined specificity. The analysis of specific antibodies from unrelated individuals indicated a virtual absence of idiotypic cross-reactivity at the level of framework idiotopes (96), whereas a remarkable degree of similarity was found when comparing idiotopes associated with the antigen-binding site. Although this was observed with specific IgG antibodies, it was particularly striking for IgE antibodies, for which cross-reactivity was found in 96% of assays run with antibodies isolated from 10 unrelated individuals. Cross-reactivity was judged significant if at least 50% binding or inhibition, depending on the assay, was seen after the addition of anti-idiotypic antibodies of the other nine individuals (98). In theory therefore, pools of immunoglobulins could be used to induce anti-idiotypic antibodies to the paratopes of antiallergen antibodies. In this case, however, the immune response of the recipient would also be directed to allotypic determinants of the immunizing molecule.

At first glance, these results are not in keeping with other studies of idiotypic cross-reactivity of specific antibodies from unrelated individuals. In the anti-TT immune response, sharing of idiotypic determinants was infrequent (18). In this study, however, rabbit antibodies were used to characterize cross-reactive idiotypes. Rabbits preferentially produce antibodies to framework-associated idiotopes (96), in contrast to auto-anti-idiotypic antibodies. In addition, it is possible that the immune response to allergens is particular with respect to idiotype expression. We have recently found that specific anti-*D. pteronyssinus* antibodies isolated by immunoadsorption cross-reacted with anti-double-stranded DNA (69a); idiotypic cross-reactivity on anti-DNA antibodies is widely recognized (70). Other investigators have found cross-reactive paratope-associated idiotopes carried by human anti-*N*-acetylglucosamine antibodies, as defined by monoclonal anti-idiotypic antibodies (122).

Whatever the explanation, our observations on anti-*D. pteronyssinus* and anti-grass pollen antibodies led us to envisage only a strictly autologous therapy.

Idiotype immunogenicity

It was believed that an acceptable means of increasing idiotype immunogenicity was to use complexes made from specific antibodies and allergen. As a rule, complexes have always been prepared in antibody excess, which should preferentially induce anti-idiotypic antibodies to framework-associated idiotopes (69). It was also reasoned that if, by any chance, inoculation of allergen-antibody complexes happened to increase allergen immunogenicity, this would nevertheless not be detrimental to patients, and could actually be beneficial, since this is the very goal of conventional hyposensitization.

Clinical Trials

Anti-idiotypic antibody production

To evaluate whether inoculation of allergen-antibody complexes could induce an increase in anti-idiotypic antibody production, four patients hypersensitive to *D. pteronyssinus* received a series of intradermal inoculations with complexes of allergen and an excess of autologous specific antibodies (73). A rapid increase in the level of anti-idiotypic antibodies was observed, which reached a plateau after a few weeks and was maintained at this level for the 12 months of the trial.

Study design

We have conducted a series of double-blind studies on patients suffering from allergic bronchial asthma related to hypersensitivity to *D. pteronyssinus* or to grass pollen. Patients were inoculated with allergen-antibody complexes made from allergen combined with autologous specific antiallergen antibodies isolated by immunoadsorption. The total amount of specific antibodies which was used for each patient did not exceed an average of 10 μg, and the quantity of allergen was at least 100-fold smaller than with conventional hyposensitization. Symptom and medication scores were monitored by using diary cards. In the study of *D. pteronyssinus* hypersensitivity, peak expiratory flows were measured daily and bronchial challenges were performed on several occasions with both allergen and acetylcholine, to evaluate specific and nonspecific bronchial responsiveness. In addition, skin reactivity was regularly evaluated by intradermal allergen inoculation, and the level of specific antibodies to allergen was monitored.

Clinical outcome

The inoculation of allergen-antibody complexes was well tolerated. Systemic reactions were never observed, and the local immediate wheal-and-flare reaction never exceeded a 2-cm diameter. In the *D. pteronyssinus* trial (74), the first signs of clinical improvement were recorded after a few weeks of treatment. Both symptom and medication scores gradually diminished, and after an average of 3 months of treatment the need for corticotherapy had completely disappeared in treated patients, whereas 60% of the placebo patients still had to use it regularly. At the end of the first year, 70% of the treated patients had lost their bronchial reactivity to allergen inhalation and exhibited a marked reduction in allergen skin reactivity. Thirty percent of the treated patients were able to discontinue their medication. These results were maintained for the following years, during which maintenance inoculations of allergen-antibody complex were given at 3-month intervals. A significant reduction of nonspecific bronchial reactivity was observed after the third year of treatment.

In the grass pollen hypersensitivity trial (72), inoculation of allergen-antibody complexes completely prevented the appearance of bronchial asthma related to pollen exposure, while 60% of placebo patients experienced regular bronchospastic reactions. No bronchodilators were used by the patients treated. A significant reduction of rhinorrhea and sneezing was noted, and none of the treated patients complained of nasal obstruction. Antihistamines were taken significantly less often by the treated group.

Evolution of specific antiallergen antibodies

In contrast to conventional hyposensitization, specific anti-*D. pteronyssinus* IgE antibody levels did not increase during the first weeks of treatment. Instead, a progressive decline in the IgE level was observed, becoming significant at week 10. After 3 months from the beginning of the treatment, the lowest values were obtained and were maintained at this level for at least 2 years. Specific IgG antibody levels followed essentially the same pattern. There was an early reduction in the IgG concentration in serum; after 6 months of treatment, mean concentrations of specific IgG increased slightly in a majority of patients, but in none of them did IgG antibodies regain their initial value. In patients with grass pollen hypersensitivity, we observed an absence of increase in levels of specific IgE antibodies during the pollen season and a progressive and significant decline afterwards. The concentration of specific IgG antibodies in serum was not modified in patients with grass pollen hypersensitivity.

Conclusions

These pilot trials indicate that inoculation of complexes made from allergen and an excess of specific autologous antibodies produces a fast and significant clinical improvement in immediate hypersensitivity to either *D. pteronyssinus* or grass pollens. The most striking effects were observed in patients with bronchial asthma, in whom reactivity to allergen inhalation disappeared. On long-term therapy, nonspecific bronchial reactivity also improved significantly. These results were accompanied by a drastic reduction in skin reactivity to allergen inoculation. The reduction in parallel of the levels of both specific IgE and IgG antibodies suggests that allergen-antibody complexes selectively suppress the production of antiallergen antibodies in vivo. Changes in the concentrations of anti-idiotypic antibodies in serum may indicate that this suppression is mediated through idiotypic suppression. Preliminary data suggest that the secondary increase in the level of specific IgG antibodies can be attributed to the emergence of antibodies carrying new idiotypic determinants. Although present, this phenomenon does not seem to be very significant for IgE antibodies. Finally, a recent clinical trial comparing the effect of inoculation of specific antibodies alone, of allergen alone, and of allergen-antibody complexes has demonstrated that these clinical and biological results cannot be obtained by either specific antibodies or allergen alone.

CONCLUSIONS

There remains little doubt that in animals, the production of isologous anti-idiotypic antibodies can be enhanced by immunization with specific antibodies. The anti-idiotypic antibodies exert an efficient suppressive activity on the formation of IgE antibodies, which appear more amenable to suppression than other isotypes owing to their low concentration, the small number of B cells committed to IgE synthesis, and the apparent incapacity to shift idiotype expression on IgE molecules. Anti-idiotypic antibodies raised to polyclonal antibodies against a protein are efficient in suppressing the production of all antibodies toward this protein. I have enumerated the reasons why care should be taken in the extrapolation of experimental results to the human situation, primarily because of the absence of animal models of spontaneous IgE production.

Circumstantial evidence for a physiological role of anti-idiotypic antibodies in the modulation of IgE formation in humans is convincing, even if direct proof of their involvement in vivo has not appeared yet. This evidence does not contradict that obtained for other diseases, such as myasthenia gravis. Nevertheless, it is difficult to accept that an inefficient control of IgE production by anti-idiotypic antibodies is in some way responsible for the "exaggerated" production of IgE antibodies in atopic individuals. An increase of the ratio of idiotype-bearing over corresponding anti-idiotypic antibodies might, however, lead to a reduction in the production of specific antiallergen IgE antibodies. To increase the immunogenicity of idiotypes, we have used complexes of allergen and autologous specific antibodies, which offer the additional advantage of favoring the production of anti-idiotypic antibodies directed to framework idiotopes. The latter are thought to be more efficient in the regulation of antibody production.

First clinical applications have demonstrated that the expected large diversity of idiotypes carried by human antiallergen antibodies does not preclude their use for anti-idiotypic antibody induction, provided that purified autologous antibodies are used. The procedure appears to be safe and of clinical benefit so far, although much work remains to be carried out to gain a better understanding of the mechanism of action. Indeed, besides their potent boosting action on the production of anti-idiotypic antibodies, antigen-antibody complexes have a number of other physiological effects on the immune system which might contribute to the clinical efficacy.

The particular susceptibility of IgE production to idiotype manipulation makes this approach worthy of further investigations, which hold promise for the improvement of the difficult situation with which the hypersensitive patient is faced.

Acknowledgments. I thank Alan Barclay, Marc Jacquemin, and Philippe Lebrun for their critical review of the manuscript and Brigitte Firket for secretarial assistance.

REFERENCES

1. **Bernstein, I. L., J. G. Michael, S. Malkiel, L. C. Sweet, and R. G. Brackett.** 1979. Immunoregulatory function of specific IgG. II. Clinical evaluation of combined active and passive immunotherapy. *Int. Arch. Allergy Appl. Immunol.* **58:**30–37.
2. **Blaser, K., M. Geiser, and A. L. de Weck.** 1979. Suppression of phosphorylcholine-spe-

cific IgE antibody formation in BALB/c mice by isologous anti-T15 antiserum. *Eur. J. Immunol.* **9**:1017–1020.

3. **Blaser, K., and A. L. de Weck.** 1982. Regulation of the IgE antibody response by idiotype-anti-idiotype network. *Prog. Allergy* **32**:203–264.

4. **Blaser, K., and C. H. Heusser.** 1984. Regulatory effects of isologous anti-idiotypic antibodies on the formation of different immunoglobulin classes in the immune response to phosphorylcholine in BALB/c mice. *Eur. J. Immunol.* **14**:93–98.

5. **Blaser, K., T. Nakagawa, and A. L. de Weck.** 1980. Suppression of the benzylpenicilloyl- (BPO) specific IgE formation with isologous anti-idiotypic antibodies in BALB/c mice. *J. Immunol.* **125**:24–30.

6. **Blaser, K., T. Nakagawa, and A. L. de Weck.** 1981. Investigation of a syngeneic murine model for the study of IgE antibody regulation with isologous anti-idiotypic antibodies. *Int. Arch. Allergy Appl. Immunol.* **64**:42–50.

7. **Blaser, K., T. Nakagawa, and A. L. de Weck.** 1981. Regulation of the IgE antibody response by anti-idiotypic antibodies and its resolution into hapten- and carrier-specific compartments. *Int. Arch. Allergy Appl. Immunol.* **66(Suppl. 1)**:61–63.

8. **Blaser, K., T. Nakagawa, and A. L. de Weck.** 1981. Suppression of anti-hapten IgE and IgG antibody responses by isologous anti-idiotypic antibodies against purified anti-carrier (ovalbumin) antibodies in BALB/c mice. *J. Immunol.* **126**:1180–1184.

9. **Blaser, K., T. Nakagawa, and A. L. de Weck.** 1983. Effect of passively administered isologous anti-idiotypes directed against anti-carrier (ovalbumin) antibodies on the anti-hapten IgE and IgG antibody responses in BALB/c mice. *Immunology* **48**:423–431.

10. **Blaser, K., A. Wetterwald, E. Weber, V. E. Gerber, and A. L. de Weck.** 1983. Immune networks in immediate type allergic diseases. *Ann. N.Y. Acad. Sci.* **418**:330–345.

11. **Bloem, A., G. Zenke, K. Eichmann, and F. Emmrich.** 1988. Human immune response to group A streptococcal carbohydrate (A-CHO). II. Antigen independent stimulation of IgM anti-A-CHO production in purified B cells by a monoclonal anti idiotopic antibody. *J. Immunol.* **140**:277–282.

12. **Bona, C. A., and A. S. Fauci.** 1980. In vitro idiotypic suppression of chronic lymphocytic leukemia lymphocytes secreting monoclonal immunoglobulin M anti-sheep erythrocyte antibody. *J. Clin. Invest.* **65**:761–767.

13. **Bona, C. A., E. Heber-Katz, and W. E. Paul.** 1981. Idiotype-anti-idiotype regulation. I. Immunization with a levan-binding myeloma protein leads to the appearance of auto-anti-(anti-idiotype) antibodies and to the activation of silent clones. *J. Exp. Med.* **153**:951–967.

14. **Bose, R., A. K. M. Ekramoddoullah, F. T. Kisil, and A. H. Sehon.** 1988. Human and murine antibodies to Rye grass pollen allergen *Lol pIV* share a common idiotope. *Immunology* **63**:579–584.

15. **Bose, R., D. G. Marsh, G. Delespesse.** 1986. Anti-idiotypes to anti-*Lol pI* (Rye) antibodies in allergic and non-allergic individuals. Influence of immunotherapy. *Clin. Exp. Immunol.* **66**:231–240.

16. **Bose, R., D. G. Marsh, J. Duchateau, A. H. Sehon, and G. Delespesse.** 1984. Demonstration of auto-anti-idiotypic antibody cross-reacting with public idiotypic determinants in the serum of Rye-sensitive allergic patients. *J. Immunol.* **133**:2474–2478.

17. **Bousquet, J., and M. D. Valentine.** 1985. Immunotherapy with hymenoptera venoms, p. 326–334. *In* C. E. Reed (ed.), *Proceedings of the XII International Congress of Allergology and Clinical Immunology.* The C. V. Mosby Co., St. Louis.

18. **Brozek, C. M. K., and R. S. Geha.** 1987. Crossreactivity and inheritance of idiotypes restricted to human anti-tetanus toxoid antibodies. *J. Clin. Invest.* **79**:1242–1248.

19. **Capron, A., J. P. Dessaint, M. Capron, and H. Bazin.** 1975. Specific IgE antibodies in

immune adherence of normal macrophages to *Schistosoma mansoni* schistosomules. *Nature* (London) **253**:474–475.

20. **Castracane, J. M., T. J. Hall, R. Tamir, and R. E. Rocklin.** 1985. Anti-ragweed-specific anti-idiotypic antibodies in nonatopic and atopic subjects. *Trans. Assoc. Am. Physicians* **98**:123–130.

21. **Castracane, J. M., and R. E. Rocklin.** 1988. Detection of human auto-anti-idiotypic antibodies (Ab2). I. Isolation and characterization of Ab2 in the serum of a ragweed immunotherapy-treated patient. *Int. Arch. Allergy Appl. Immunol.* **86**:288–294.

22. **Castracane, J. M., and R. E. Rocklin.** 1988. Detection of human auto-anti-idiotypic antibodies (Ab2). II. Generation of Ab2 in atopic patients undergoing allergen immunotherapy. *Int. Arch. Allergy Appl. Immunol.* **86**:295–302.

23. **Cerny, J., and G. Kelsoe.** 1984. Priority of the anti-idiotypic response after antigen administration: artefact or intriguing network mechanism? *Immunol. Today* **5**:61–63.

24. **Chang, S. P., R. M. Perlmutter, M. Brown, C. H. Heusser, L. Hood, and M. B. Rittenberg.** 1984. Immunologic memory to phosphocholine. IV. Hybridomas representative of group I (T15-like) and group II (non-T15-like) antibodies utilize distinct V_H genes. *J. Immunol.* **132**:1550–1555.

25. **Chiorazzi, N., D. A. Fox, and D. H. Katz.** 1976. Hapten-specific IgE antibody responses in mice. VI. Selective enhancement of IgE antibody production by low doses of X-irradiation and by cyclophosphamide. *J. Immunol.* **117**:1629–1637.

26. **Chua, K. Y., G. A. Stewart, W. R. Thomas, R. J. Simpson, R. J. Dilworth, T. M. Plozza, and K. J. Turner.** 1988. Sequence analysis of cDNA coding for a major house dust mite allergen, *Der pI*. Homology with cysteine proteases. *J. Exp. Med.* **167**:175–182.

27. **Coffman, R. L., J. Ohara, M. W. Bond, J. Carty, A. Zlotnik, and W. E. Paul.** 1986. B cell stimulatory factor-1 enhances the IgE response of lipopolysaccharide-activated B cells. *J. Immunol.* **136**:4538–4541.

28. **Couraud, P. O., B. Z. Lü, and A. D. Strosberg.** 1983. Cyclical antiidiotypic response to anti-hormone antibodies due to neutralization by autologous anti-antiidiotype antibodies that bind hormone. *J. Exp. Med.* **157**:1369–1378.

29. **Creticos, P. S., and P. S. Norman.** 1987. Immunotherapy with allergens. *J. Am. Med. Assoc.* **258**:2874–2880.

30. **Dessein, A., R. N. Germain, M. E. Dorf, and B. Benacerraf.** 1979. IgE responses to synthetic polypeptide antigens. I. Simultaneous *Ir* gene and isotype-specific regulation of IgE responses to L-glutamic acid-L-alanine-L-tyrosine (GAT). *J. Immunol.* **123**:463–470.

31. **Dessein, A., S. T. Ju, M. E. Dorf, B. Benacerraf, and R. N. Germain.** 1980. IgE response to synthetic polypeptide antigens. II. Idiotypic analysis of the IgE response to L-glutamic acid-L-alanine-L-tyrosine. *J. Immunol.* **124**:71–76.

32. **Dy, M., E. Schneider, L. N. Gastinel, C. Auffray, J. J. Mermod, and J. Hamburger.** 1987. Histamine-producing cell-stimulating activity. A biological activity shared by interleukin 3 and granulocyte-macrophage colony-stimulating factor. *Eur. J. Immunol.* **17**:1243–1248.

33. **Eichmann, K.** 1974. Idiotype suppression. I. Influence of the dose and the effector functions of anti-idiotypic antibody on the production of an idiotype. *Eur. J. Immunol.* **4**:296–302.

34. **Eichmann, K.** 1975. Idiotype suppression. II. Amplification of a suppressor T cell with anti-idiotypic activity. *Eur. J. Immunol.* **5**:511–517.

35. **Eichmann, K., A. Coutinho, and F. Melchers.** 1977. Absolute frequencies of lipopolysaccharide-reactive B cells producing A5A idiotype in unprimed, streptococcal A carbohydrate-primed, anti-A5A idiotype-sensitized and anti-A5A idiotype-suppressed A/J mice. *J. Exp. Med.* **146**:1436–1449.

36. **Eichmann, K., and K. Rajewsky.** 1975. Induction of T and B cell immunity by anti-idiotypic antibody. *Eur. J. Immunol.* **5:**661–666.
37. **Emmrich, F., G. Zenke, and K. Eichmann.** 1986. Isotype restriction of idiotopes associated with human anti-streptococcal A carbohydrate antibodies. *Eur. J. Immunol.* **16:**542–546.
38. **Fink, J. N.** 1985. Immunotherapy of asthma. *J. Allergy Clin. Immunol.* **76:**402–404.
39. **Fong, S., T. A. Gilbertson, P. P. Chen, J. H. Vaughan, and D. A. Carson.** 1984. Modulation of human rheumatoid factor-specific lymphocyte responses with a cross-reactive anti-idiotype bearing the internal image of antigen. *J. Immunol.* **132:**1183–1189.
40. **Gearhart, P. J., N. D. Johnson, R. Douglas, and L. Hood.** 1981. IgG antibodies to phosphorylcholine exhibit more diversity than their IgM counterparts. *Nature* (London) **291:**29–34.
41. **Geczy, A. F.** 1977. Histocompatibility antigens and genetic control of the immune response in guinea pigs. IV. Specific inhibition of lymphocyte proliferation by auto-anti-idiotypic antibodies. *J. Exp. Med.* **145:**1093–1098.
42. **Geczy, A. F., and A. L. de Weck.** 1974. Histocompatibility antigens and genetic control of the immune response in guinea pigs. I. Specific inhibition of antigen-induced lymphocyte stimulation by alloantisera. *Eur. J. Immunol.* **4:**483–489.
43. **Geczy, A. F., C. L. Geczy, and A. L. de Weck.** 1975. Histocompatibility antigens and genetic control of the immune response in guinea pigs. II. Specific inhibition of antigen-induced lymphocyte proliferation by anti-receptor alloantisera. *Eur. J. Immunol.* **5:**711–719.
44. **Geczy, A. F., C. L. Geczy, and A. L. de Weck.** 1976. Histocompatibility antigens and genetic control of the immune response in guinea pigs. III. Specific inhibition of antigen-induced lymphocyte proliferation by strain-specific anti-idiotypic antibodies. *J. Exp. Med.* **144:**226–240.
45. **Geczy, A. F., A. L. de Weck, C. L. Geczy, and O. Toffler.** 1978. Suppression of reaginic antibody formation in guinea pigs by anti-idiotypic antibodies. *J. Allergy Clin. Immunol.* **62:**261–270.
46. **Geha, R. S.** 1982. Elicitation of the Prausnitz-Küstner reaction by antiidiotypic antibodies. *J. Clin. Invest.* **69:**735–741.
47. **Geha, R. S.** 1982. Presence of auto-anti-idiotypic antibody during the normal human immune response to tetanus toxoid antigen. *J. Immunol.* **129:**139–144.
48. **Geha, R. S.** 1983. Presence of circulating anti-idiotype-bearing cells after booster immunization with tetanus toxoid (TT) and inhibition of anti-TT antibody synthesis by auto-anti-idiotypic antibody. *J. Immunol.* **130:**1634–1639.
49. **Geha, R. S., and M. Comunale.** 1983. Regulation of immunoglobulin E antibody synthesis in man by antiidiotypic antibodies. *J. Clin. Invest.* **71:**46–54.
50. **Geha, R. S., and R. P. Weinberg.** 1978. Anti-idiotypic antisera in man. I. Production and immunochemical characterization of anti-idiotypic antisera to human antitetanus antibodies. *J. Immunol.* **121:**1518–1523.
51. **Gill-Pazaris, L. A., E. Lamoyi, A. R. Brown, and A. Nisonoff.** 1981. Properties of a minor cross-reactive idiotype associated with anti-p-azophenylarsonate antibodies of A/J mice. *J. Immunol.* **126:**75–79.
52. **Gilles, J. G. G., J. C. P. Mareschal, and J. M. R. Saint-Remy.** 1988. Latex allergosorbent test (LAST): a new immunoassay for specific IgE with latex particles. *J. Allergy Clin. Immunol.* **82:**35–39.
53. **Gleich, G. J., and R. T. Jones.** 1975. Measurement of IgE antibodies by the radioallergosorbent test. I. Technical considerations in the performance of the test. *J. Allergy Clin. Immunol.* **55:**334–345.
54. **Goldberg, B., W. E. Paul, and C. A. Bona.** 1983. Idiotype-antiidiotype regulation. IV.

Expression of common regulatory idiotopes on fructosan-binding and non-fructosan-binding monoclonal immunoglobulin. *J. Exp. Med.* 8:515–528.

55. **Haba, S., E. M. Rosen, K. Meek, and A. Nisonoff.** 1986. Primary structure of IgE monoclonal antibodies expressing an intrastrain crossreactive idiotype. *J. Exp. Med.* **164**:291–302.

56. **Hamaoka, T., D. H. Katz, K. J. Bloch, and B. Benacerraf.** 1973. Hapten-specific IgE antibody responses in mice. I. Secondary IgE responses in irradiated recipients of syngeneic primed spleen cells. *J. Exp. Med.* **138**:306–311.

57. **Hamaoka, T., D. H. Katz, K. J. Bloch, and B. Benacerraf.** 1973. Hapten-specific IgE antibody responses in mice. II. Cooperative interactions between adoptively transferred T and B lymphocytes in the development of IgE response. *J. Exp. Med.* **138**:538–556.

58. **Hamburger, R. N.** 1979. Inhibition of IgE binding to tissue culture cells and leucocytes by pentapeptide. *Immunology* **38**:781–787.

59. **Hamilton, R. G., A. K. Sobotka, and N. F. Adkinson.** 1979. Solid-phase radioimmunoassay for quantitation of antigen-specific IgG in human serum with ^{125}I-protein A from Staphylococcus aureus. *J. Immunol.* **122**:1073–1079.

60. **Heyman, B., and H. Wigzell.** 1984. Immunoregulation by monoclonal sheep erythrocyte-specific IgG antibodies: suppression is correlated to level of antigen binding and not to isotype. *J. Immunol.* **132**:1136–1143.

61. **Hiernaux, J. R., J. Chiang, P. J. Baker, C. Delisi, and B. Prescott.** 1982. Lack of involvement of auto-anti-idiotypic antibody in the regulation of oscillations and tolerance in the antibody response to levan. *Cell. Immunol.* **67**:334–345.

62. **Hiernaux, J. R., J. Marvel, P. Meyers, M. Moser, O. Leo, M. Slaoui, and J. Urbain.** 1986. Study of idiotopic suppression induced by anti-cross-reactive idiotype monoclonal antibody in the anti-*p*-azophenylarsonate antibody response. *J. Immunol.* **136**:1960–1967.

63. **Hirai, Y., E. Lamoyi, Y. Dohi, and A. Nisonoff.** 1981. Regulation of expression of a family of cross-reactive idiotypes. *J. Immunol.* **126**:71–74.

64. **Hirano, T., S. Kojima, N. Hirayama, M. J. Nelles, T. Indada, A. Nisonoff, and Z. Ovary.** 1983. Presence of an intrastrain cross-reactive idiotype on A/J antibodies of the IgE class specific for the *p*-azophenylarsonate group. *J. Immunol.* **130**:1300–1302.

64a.**Jacquemin, M. G., and J. M. R. Saint-Remy.** Anti-idiotypic antibodies in the immune response of nonatopic subjects toward allergens. Manuscript in preparation.

65. **Jarrett, E. E. E.** 1978. Stimuli for the production and control of IgE in rats. *Immunol. Rev.* **41**:52–76.

66. **Jerne, N. K.** 1984. Idiotypic networks and other preconceived ideas. *Immunol. Rev.* **79**:5–24.

67. **Jerne, N. K., J. Roland, and P. A. Cazenave.** 1982. Recurrent idiotopes and internal images. *EMBO J.* **1**:243–247.

68. **Kelsoe, G., M. Reth, and K. Rajewsky.** 1980. Control of idiotope expression by monoclonal anti-idiotope antibodies. *Immunol. Rev.* **52**:75–88.

69. **Klaus, G. G. B.** 1978. Antigen-antibody complexes elicit anti-idiotypic antibodies to self-idiotopes. *Nature* (London) **272**:265–266.

69a.**Lebrun, P. M., M. G. Jacquemin, and J. M. R. Saint-Remy.** Antibodies to *D. pteronyssinus* allergens cross-react with ssDNA. Manuscript in preparation.

70. **Livneh, A., J. L. Preud'homme, A. Solomon, and B. Diamond.** 1987. Preferential expression of the systemic lupus erythematosus-associated idiotype 8.12 in sera containing monoclonal immunoglobulins. *J. Immunol.* **139**:3730–3733.

71. **Lopez, A. F., C. J. Sanderson, J. R. Gamble, H. D. Campbell, I. G. Young, and M. A. Vadas.** 1988. Recombinant human interleukin 5 is a selective activator of human eosinophil function. *J. Exp. Med.* **167**:219–224.

72. **Machiels, J. J., M. A. Somville, M. G. Jacquemin, and J. M. R. Saint-Remy.** 1990. Allergen-antibody complexes can efficiently prevent seasonal rhinitis and asthma in grass-pollen hypersensitive patients. *J. Allergy Clin. Immunol.,* in press.

73. **Machiels, J. J., S. J. Lebecque, M. G. Jacquemin, and J. M. R. Saint-Remy.** 1990. Antigen-antibody complexes can increase the level of specific anti-idiotypic antibodies in patients hypersensitive to Dermatophagoides pteronyssinus. Submitted for publication.

74. **Machiels, J. J., M. A. Somville, P. M. Lebrun, S. J. Lebecque, M. G. Jacquemin, and J. M. R. Saint-Remy.** 1990. Allergic bronchial asthma due to *D. pteronyssinus* hypersensitivity can be efficiently treated by inoculation of allergen-antibody complexes. *J. Clin. Invest.,* in press.

75. **MacKenzie, T., and H. M. Dosch.** 1989. Clonal and molecular characteristics of the human IgE-committed B cell subset. *J. Exp. Med.* **169:**407–430.

76. **Malipiero, U. V., N. S. Levy, and P. J. Gearhart.** 1987. Somatic mutation in anti-phosphorylcholine antibodies. *Immunol. Rev.* **96:**59–74.

77. **Malley, A., L. M. Bradley, and S. M. Shiigi.** 1987. The role of macrophages in anti-idiotypic antibody and T suppressor factor induction of Timothy grass pollen antigen B-specific T suppressor cells. *J. Immunol.* **139:**1046–1053.

78. **Malley, A., C. J. Brandt, and L. B. Deppe.** 1982. Preparation and characterization of the anti-idiotypic properties of rabbit anti-Timothy antigen B helper factor and anti-mouse Timothy IgE antisera. *Immunology* **45:**217–225.

79. **Malley, A., and D. W. Dresser.** 1982. Anti-idiotype regulation of Timothy grass pollen IgE antibody formation. I. *In vivo* induction of suppressor T cells. *Immunology* **46:**653–659.

80. **Malley, A., and D. W. Dresser.** 1983. Anti-idiotype regulation of the formation of IgE antibody to Timothy grass pollen. II. *In vitro* induction of suppressor T cells in mini-Marbrook cultures. *Immunology* **48:**93–99.

81. **Marsh, D. G., S. H. Hsu, M. Roebber, E. Ehrlich-Kautzky, L. R. Freidhoff, D. A. Meyers, M. K. Pollard, and W. B. Bias.** 1982. HLA-Dw2: a genetic marker for human immune response to short ragweed pollen allergen Ra5. I. Response resulting primarily from natural antigenic exposure. *J. Exp. Med.* **155:**1439–1451.

82. **Mecheri, S., W. Mourad, J. Lapeyre, M. Jobin, B. David, and J. Hébert.** 1988. Demonstration of idiotypes expressed on basophil-bound IgE antibodies by using anti-idiotype-induced histamine release in grass pollen-allergic patients. *Immunology* **64:**11–15.

83. **Meek, K., C. Hasemann, B. Pollock, S. S. Alkan, M. Brait, M. Slaoui, J. Urbain, and J. D. Capra.** 1989. Structural characterization of antiidiotypic antibodies. Evidence that Ab2s are derived from the germline differently than Ab1s. *J. Exp. Med.* **169:**519–533.

84. **Mourad, W., G. Pelletier, and J. Hébert.** 1988. Characterization of anti-idiotypic antibodies and their use as probes for determination of shared idiotopes expressed on murine and human IgE anti-Rye I antibodies. *Immunology* **63:**397–401.

85. **Mudawwar, F., Z. Awdeh, K. Ault, and R. S. Geha.** 1980. Regulation of monoclonal immunoglobulin G synthesis by antiidiotypic antibody in a patient with hypogammaglobulinemia. *J. Clin. Invest.* **65:**1202–1209.

86. **Müller, C. E., and K. Rajewsky.** 1984. Idiotope regulation by isotype switch variants of two monoclonal antiidiotope antibodies. *J. Exp. Med.* **159:**758–772.

87. **Nakagawa, T., K. Blaser, and A. L. de Weck.** 1980. Rat mast cell degranulation triggered by murine anti-idiotypic antibodies. *Int. Arch. Allergy Appl. Immunol.* **63:**462–466.

88. **Nemazee, D. A., and V. L. Sato.** 1982. Enhancing antibody: a novel component of the immune response. *Proc. Natl. Acad. Sci. USA* **79:**3828–3832.

89. **Patterson, R., M. Rosenberg, and M. Roberts.** 1977. Evidence that *Aspergillus fumigatus* growing in the airway of man can be a potent stimulus of specific and nonspecific IgE formation. *Am. J. Med.* **63:**257–262.

90. **Pène, J., F. Rousset, F. Brière, I. Chrétien, J. Wideman, J. Y. Bonnefoy, and J. E. De Vries.** 1988. Interleukin 5 enhances interleukin 4-induced IgE production by normal human B cells. The role of soluble CD23 antigen. *Eur. J. Immunol.* **18:**929–935.

91. **Platts-Mills, T. A. E., P. W. Heymann, M. D. Chapman, and E. B. Mitchell.** 1985. Bronchial hyper-reactivity and allergen exposure. *Prog. Respir. Res.* **19:**276–284.

92. **Reth, M., G. Kelsoe, and K. Rajewsky.** 1981. Idiotypic regulation by isologous monoclonal anti-idiotope antibodies. *Nature* (London) **290:**257–259.

93. **Rose, L. M., M. Goldman, and P. H. Lambert.** 1982. Simultaneous induction of an idiotype, corresponding anti-idiotypic antibodies, and immune complexes during African trypanosomiasis in mice. *J. Immunol.* **128:**79–85.

94. **Rossi, F., Y. Sultan, and M. D. Kazatchkine.** 1988. Anti-idiotypes against autoantibodies and alloantibodies to VIII:C (anti-haemophilic factor) are present in therapeutic polyspecific normal immunoglobulins. *Clin. Exp. Immunol.* **74:**311–316.

95. **Sacks, D. L., K. M. Esser, and A. Sher.** 1982. Immunization of mice against African trypanosomiasis using anti-idiotypic antibodies. *J. Exp. Med.* **155:**1108–1119.

96. **Saint-Remy, J. M. R., S. J. Lebecque, and P. M. Lebrun.** 1988. Human immune response to allergens of house dust mite, *Dermatophagoides pteronyssinus.* III. Cross-reactivity of bystander idiotopes on allergen-specific IgE antibodies. *Eur. J. Immunol.* **18:**77–81.

97. **Saint-Remy, J. M. R., S. J. Lebecque, P. M. Lebrun, and M. G. Jacquemin.** 1988. Human immune response to allergens of house dust mite, *Dermatophagoides pteronyssinus.* IV. Occurrence of natural autologous anti-idiotypic antibodies. *Eur. J. Immunol.* **18:**83–87.

98. **Saint-Remy, J. M. R., S. J. Lebecque, P. M. Lebrun, and M. G. Jacquemin.** 1988. Human immune response to allergens of house dust mite, *Dermatophagoides pteronyssinus.* V. Auto-anti-idiotypic antibody characterization and cross-reactivity. *Eur. J. Immunol.* **18:**1009–1014.

99. **Saint-Remy, J. M. R., P. M. Lebrun, S. J. Lebecque, and P. L. Masson.** 1986. The human immune response against major allergens from house dust mite, *Dermatophagoides pteronyssinus.* II. Idiotypic cross-reactions of allergen-specific antibodies. *Eur. J. Immunol.* **16:**575–580.

100. **Saint-Remy, J. M. R., P. M. Lebrun, S. J. Lebecque, and P. L. Masson.** 1988. Human immune response to allergens of house dust mite, *Dermatophagoides pteronyssinus.* Isotypic analysis of antibodies in atopic and non-atopic subjects. *Allergy* **43:**338–347.

101. **Sakato, N., and H. N. Eisen.** 1975. Antibodies to idiotypes of isologous immunoglobulins. *J. Exp. Med.* **141:**1411–1426.

102. **Sakato, N., M. Semma, H. N. Eisen, and T. Azuma.** 1982. A small hypervariable segment in the variable domain of an immunoglobulin light chain stimulates formation of anti-idiotypic suppressor T cells. *Proc. Natl. Acad. Sci. USA* **79:**5396–5400.

103. **Saxon, A., M. Kaplan, and R. H. Stevens.** 1980. Isotype-specific human B lymphocytes that produce immunoglobulin E in vitro when stimulated by pokeweed mitogen. *J. Allergy Clin. Immunol.* **66:**233–241.

104. **Schopfer, K., K. Baerlocher, P. Price, U. Krech, P. G. Quie, and S. D. Douglas.** 1979. Staphylococcal IgE antibodies, hyperimmunoglobulinemia E and *Staphylococcus aureus* infections. *N. Engl. J. Med.* **300:**835–838.

105. **Sedgwick, J. D., and P. G. Holt.** 1984. Suppression of IgE responses in inbred rats by repeated respiratory tract exposure to antigen: responder phenotype influences isotype specificity of induced tolerance. *Eur. J. Immunol.* **14:**893–897.

106. **Snapper, C. M., and W. E. Paul.** 1987. Interferon-γ and B cell stimulatory factor-1 reciprocally regulate Ig isotype production. *Science* **236:**944–947.

107. **Somville, M. A., J. Machiels, J. G. G. Gilles, and J. M. R. Saint-Remy.** 1989. Seasonal variation in specific IgE antibodies of grass-pollen hypersensitive patients depends on the

steady state IgE concentration and is not related to clinical symptoms. *J. Allergy Clin. Immunol.* **83**:486–494.

108. **Tada, T., and K. Ishizaka.** 1970. Distribution of γE-forming cells in lymphoid tissues of the human and monkey. *J. Immunol.* **104**:377–387.

109. **Takatsu, K., and K. Ishizaka.** 1976. Reaginic antibody formation in the mouse. VII. Induction of suppressor T cells for IgE and IgG antibody responses. *J. Immunol.* **116**:1257–1264.

110. **Tasiaux, N., R. Leuwenkroon, C. Bruyns, and J. Urbain.** 1978. Possible occurrence and meaning of lymphocytes bearing autoanti-idiotypic receptors during the immune response. *Eur. J. Immunol.* **8**:464–468.

111. **Taylor, R. B., J. P. Tite, and C. Manzo.** 1979. Immunoregulatory effects of a covalent antigen-antibody complex. *Nature* (London) **281**:488–490.

112. **Trenkner, E., and R. Riblet.** 1975. Induction of antiphosphorylcholine antibody formation by anti-idiotypic antibodies. *J. Exp. Med.* **142**:1121–1132.

113. **Tung, A. S., N. Chiorazzi, and D. H. Katz.** 1978. Regulation of IgE antibody production by serum molecules. I. Serum from complete Freund's adjuvant-immune donors suppresses irradiation-enhanced IgE production in low responder mouse strains. *J. Immunol.* **120**:2050–2059.

114. **Vaz, N. M., and B. B. Levine.** 1970. Immune responses of inbred mice to repeated low doses of antigen: relationship to histocompatibility (H-2) type. *Science* **168**:852–854.

115. **Vercelli, D., and R. S. Geha.** 1989. Regulation of IgE synthesis in humans. *J. Clin. Immunol.* **9**:75–83.

116. **Weber, E. A., C. H. Heusser, and K. Blaser.** 1986. Fine specificity and idiotype expression of anti-phosphorylcholine IgE and IgG antibodies. *Eur. J. Immunol.* **16**:919–923.

117. **Weissberger, H. Z., R. R. Shenk, and H. B. Dickler.** 1983. Antiidiotype stimulation of antigen-specific antigen-independent antibody responses in vitro. I. Evidence for stimulation of helper T lymphocyte function. *J. Exp. Med.* **158**:465–476.

118. **Wetterwald, A., O. Toffler, A. L. de Weck, and K. Blaser.** 1986. Suppression of established IgE antibody responses with isologous anti-idiotypic antibodies in guinea pigs. *Mol. Immunol.* **23**:347–356.

119. **Wikler, M., J. D. Franssen, C. Collignon, O. Leo, B. Mariamé, P. Van de Walle, D. De Groote, and J. Urbain.** 1979. Idiotypic regulation of the immune system. Common idiotypic specificities between idiotypes and antibodies raised against anti-idiotypic antibodies in rabbits. *J. Exp. Med.* **150**:184–195.

120. **Wilkinson, I., C. J. C. Jackson, G. M. Lang, V. Holford-Strevens, and A. H. Sehon.** 1987. Tolerance induction in mice by conjugates of monoclonal immunoglobulins and monomethoxypolyethylene glycol. Transfer of tolerance by T cells and by T cell extracts. *J. Immunol.* **139**:326–331.

121. **Yaoita, Y., Y. Kumagai, K. Okumura, and T. Honjo.** 1982. Expression of lymphocyte surface IgE does not require switch recombination. *Nature* (London) **297**:697–699.

122. **Zenke, G., K. Eichmann, and F. Emmrich.** 1984. Characterization of a major human antibody clonotype (1A) by monoclonal antibodies to combining site-associated idiotopes. *Eur. J. Immunol.* **14**:164–170.

Chapter 8

Idiotypy in Autoimmunity

*Maurizio Zanetti, Nebojsa Dovezenski, Petar Lenert,
and Maurizio Sollazzo*

Autoimmune diseases are conditions in which structural or functional damage to self constituents (autoantigens) occurs as a result of humoral or cell-mediated immune reactions. They can be classified into organ-specific (if only one organ is involved) and systemic (if various organs are affected) diseases. Autoimmune diseases can be caused by, or be associated with, autoantibodies to structural moieties of the body tissues and cells (basement membrane, microsomal antigens, DNA, cell surface receptors, etc.) or their products (spermatozoa, thyroglobulin, etc.).

At the beginning of this century, Ehrlich and Morgenroth (29) observed that animals do not generate antibodies against their own tissues and interpreted this phenomenon, which they termed horror autotoxicus, as evidence that the body does not have cells with receptors directed against self tissues. This has become a central dogma in immunology, with a long-standing influence on our views on autoimmunity. In recent years, however, it has become apparent that responses involving self components exist during the so-called normal state, that cells of normal adult animals bear receptors for autoantigens (30, 116), and that polyclonal activators (54) can readily induce the production of autoantibodies. While this implies that within the pool of quiescent lymphocytes there are those precommitted to self reactivity, it has created the necessity to revise the old paradigm.

Studies at the level of the neonatal preimmune B-cell repertoire in normal animals have further elucidated that recognition of self antigens is prevalent (39). Both the emerging immune system and that of adults appear

Maurizio Zanetti, Nebojsa Dovezenski, Petar Lenert, and Maurizio Sollazzo • Division of Dermatology, Department of Medicine, University of California, San Diego, 225 Dickinson Street, San Diego, California 92103.

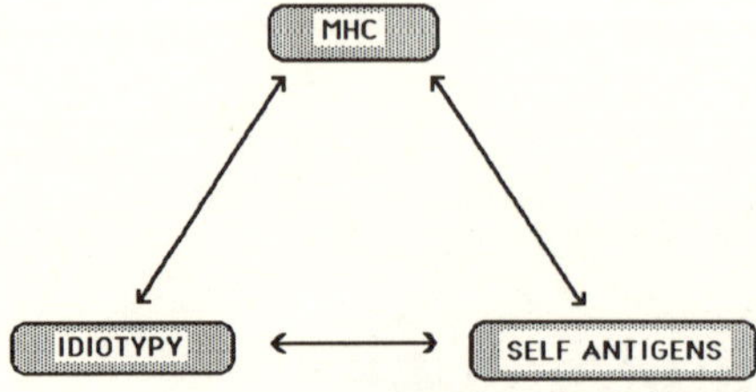

Figure 1. Molecular determinants participating in the self-recognition process.

to be programmed to respond to self components, implying that this may be not only a constitutive function of the immune system but also a safe physiological process. In addition, several studies now suggest (48) that interaction among constitutive structures of the internal milieu, e.g., members of the immunoglobulin gene superfamily, may also have an important physiological role.

Three categories of self moieties are believed to participate in the self-recognition process (Fig. 1): polymorphic major histocompatibility complex determinants; nonpolymorphic organ-specific antigens, i.e., classical autoantigens; and antigenic determinants of the variable (V) region of the antibody molecules, idiotypes. Self-recognition is not unique to the immune system. Both the endocrine and nervous systems also base their activities on such a modus operandi. For the sake of this review, we will concentrate on idiotypy of self-reactive antibodies.

Idiotypy was discovered in 1963 by Oudin and Michel (83) and Kunkel et al. (62). Much evidence has accumulated since to indicate that idiotypes are central to regulation of the immune system, i.e., the idiotype network (56). The prerequisites for a physiological regulation of the immune system based on idiotypy, endogenous regulation, can be summarized as follows: (i) antibodies (Ab_1) produced in response to a given antigen elicit anti-idiotype antibodies (Ab_2) within the same individual (90); (ii) antibodies of the various classes of immunoglobulins in response to the same antigen share idiotypy (82); (iii) spontaneously arising anti-idiotypic antibodies have been documented in several instances (for a review, see reference 121); and (iv) the mutual recognition of idiotype and complementary anti-idiotype occurs in both free molecules and cellular receptors on B and T lymphocytes (for a review, see reference 9).

Whereas the idiotypic network may account for the maintenance of immunological homeostasis, the interplay of idiotypes and anti-idiotypes borne on self-recognizing immunoglobulin and T-cell receptor V regions is of considerable conceptual and practical importance in understanding how the immune response to autoantigens is regulated. This is directly pertinent to understanding the causes of dysregulation of the internal mechanism of control and prevention of autoimmune disease.

In this article we review the current aspects of idiotypy of autoantibodies, the genes that may be involved in their expression, the structural requirements for idiotype expression, and their methods of analysis. In addition, we also review experiments that illustrate how idiotypy can positively or negatively affect the regulation of autoimmunity.

IDIOTYPES OF AUTOANTIBODIES

Antibodies to idiotypic markers of the V region of antibody molecules have been instrumental in studies of the relatedness among molecules with specificity for the same antigen. This approach has been extensively used to analyze idiotypy of autoantibodies in both human and animal systems. We will briefly review these findings.

CRI on Autoantibodies

Idiotypic cross-reactivity has been documented for autoantibodies associated with both systemic and organ-specific autoimmune diseases. In the first group are rheumatoid factors (RFs), cold agglutinins, autoantibodies to DNA, Sm and ribonucleoprotein (RNP), and antibodies to centromere. In the second group, cross-reactive idiotypes (CRI) have been described for antibodies to the acetylcholine (ACh) receptor, thyroglobulin, insulin, tubular basement membrane antigen of the kidney, immunoglobulins of the cerebrospinal fluid in patients with multiple sclerosis, and the peripheral-nerve glycolipid antigen associated with autoimmune neuropathy. This information is summarized in Table 1.

CRI and Clinical Activity of the Disease

A correlation between the level of circulating CRI and the activity of disease in humans has been documented in several systems. In this section we will consider the variations of the CRI of anti-DNA antibodies in patients with systemic lupus erythematosus (SLE), anti-ACh receptor antibodies in patients with myasthenia gravis, and anticentromere antibodies in patients with scleroderma.

Anti-DNA antibodies

In patients with SLE, anti-DNA antibodies display a large degree of idiotypic cross-reactivity. In a study by Shoenfeld et al. (102), polyclonal anti-idiotypic antibodies to individual monoclonal autoantibodies derived from lymphocytes of two patients with SLE showed extensive idiotype sharing among 60 human anti-DNA monoclonal autoantibodies from seven SLE patients. One idiotype, 16/6, was detected in the sera of about 50% of

Table 1. CRI on Autoantibodies

Disease type	Autoantibody specificity	Clinical association	References	
			Rodent studies	Human studies
Systemic	IgG Fc	Rheumatoid arthritis	11	21, 35, 49, 61, 84, 115
	Erythrocytes	Cold agglutinin anemia	17	32, 61, 117
	DNA	SLE	3, 43, 69, 88, 112	102, 107
	Sm/RNP	SLE		86, 87
	Ro/SSA	SLE/Sjögren's syndrome		37
	Centromere	Scleroderma		46
	ACh receptor	Myasthenia gravis	97	65, 66
Organ specific	Thyroglobulin	Chronic thyroiditis	118, 119	68, 70, 123
	Insulin	Diabetes		110
	Myelin basic protein	Multiple sclerosis	36	5, 28, 108
	Peripheral-nerve glycolipid	Neuropathy		33
	Kidney TBM	Interstitial nephritis	13, 77	
	Retinal S antigen	Uveoretinitis	25	

patients with clinically active disease, and the level of its expression seems to parallel disease activity (52). The 16/6 idiotype could be detected in the glomeruli of patients with lupus nephritis (51), as well as at the dermal-epidermal junction of patients with both SLE and discoid lupus (50). In another study by Kalunian et al. (57) it was found that an idiotype (idiotype GN2) accounts for the majority of anti-DNA antibodies deposited in the glomeruli of patients with SLE nephritis. It appears, however, that none of the above-mentioned idiotypes can be used as a marker of disease, or predictor of its severity, since they can be detected on antibodies lacking DNA binding and in the course of other clinical conditions.

Anti-ACh receptor antibodies

Lefvert et al. (66) prepared several monoclonal antibodies that bind specifically to idiotypes of anti-ACh receptor antibodies in patients with myasthenia gravis. The proportion of patients expressing two idiotopes,

AI21 and AI3, was higher in severe disease (67 and 59%, respectively) than in mild disease (33 and 22%, respectively). Although the concentration of these and other idiotopes is not constant during the course of the disease, in general it seems that the concentrations of antireceptor antibody activity and idiotype correlate well. At least in some situations, anti-idiotypes show an inverse relation. During the development of the disease in patients with early myasthenia gravis and in patients who could be monitored before the start of symptoms, an increase in the number of immunoglobulin G (IgG) receptor antibodies and a decrease in the number of anti-idiotypes was noticed.

Anticentromere antibodies

Hildebrandt et al. (46) recently reported the results of a long-term longitudinal study on the expression of the idiotype of anticentromere auto-antibodies that develop in patients with scleroderma. They found that over a 13-year period there was a close correlation between the appearance and levels of anticentromere autoantibodies and the level of the associated idiotype.

ORGANIZATION AND STRUCTURE OF V GENES ENCODING AUTOANTIBODIES

The relationship between autoantibodies and antibodies directed against exogenous antigens has been the subject of numerous investigations. Studies of murine models aimed at investigating the characteristics of V genes encoding autoantibodies have provided new insights into several issues such as the organization of V genes, the utilization of various V-gene families, and the polymorphism of these genes in strains prone to autoimmune diseases. Since these are important prerequisites for the understanding of the origin of autoantibodies and their idiotypes, we will briefly review the current available information on the organization and number of immunoglobulin V genes in the mouse and in humans.

In the mouse there are approximately 200 to 1,500 V_H, 12 D, and 4 J_H gene segments. Similarly, there are about 300 V_L and 4 J_L gene segments. V_H genes, for which most information presently exists, are grouped on the basis of sequence homology into 11 clustered families (12). To date, it has been almost impossible to identify a particular gene(s) unique to autoantibodies. Perhaps the only defined gene abnormality is that of a V_κ gene in MRL and C57BL/6 *tsk* mice (58).

In humans, seven different V_H gene families have been characterized (7, 47, 64, 101). Relatively small families (1 to 10 germ line members) have been described, in addition to the 20 to 60 germ line genes of the previously characterized $V_H I$, $V_H II$, and $V_H III$ subgroups. It appears, however, that many of the antibodies derived from either Epstein-Barr virus (EBV)-transformed peripheral-blood lymphocytes or human-human fusions use V_H members of these relatively small gene families. Therefore, the human immunoglobulin V_H locus is less polymorphic than the murine one, and at least in the case of IgM autoantibodies, the V_H and V_L gene segments appear to be direct copies of germ line genes.

It was initially established (60) that in mice with murine lupus, autoantibodies to DNA derive from the same (or overlapping) set of genes as antibodies to exogenous antigens. Studies of rheumatoid factors (75, 93–95), anti-DNA (111), anti-bromelain-treated autologous erythrocytes (45, 89), and murine antithyroglobulin (105) antibodies provide evidence that there are no fundamental differences between autoantibodies that arise in various murine models of autoimmunity and autoantibodies that can be induced and/or selected in normal mice. Similarly, restriction fragment length polymorphism analysis of germ line V genes of autoimmune strains shows only subtle differences in V_H and V_L gene families between some autoimmune and normal mice strains of the same Igh haplotype (for a review, see reference 10). Although genetic restriction among immunoglobulins of a given specificity cannot be excluded, it appears that autoantibodies can be encoded by members of all known V_H genes. On the other hand, since somatic mutation has also been documented, positive selection by antigen may also be taken into account.

It has been more difficult to address these issues in humans, largely because of the difficulty in obtaining a monoclonal source of autoantibodies. Nevertheless, even in the case of human autoantibodies it appears that germ line V genes are used with little if any modification. For instance, human RFs of the Wa CRI (see below) use a germ line gene designated $V_\kappa 325$ (18), whereas paraproteins of another CRI group (Po) seem to use a gene designated $V_\kappa 328$, which encodes $V_\kappa IIIa$ subgroup L chains (22). Interestingly, the $V_\kappa 325$ gene is used in 25% of B cells in patients with chronic lymphocytic leukemia (19, 59) and frequently by polyclonal RFs in patients with Sjögren syndrome and RFs generated in vitro by EBV transformation (103).

Notably, an absolute identity has been found between the V_H region of two human anti-DNA antibodies, one derived from a patient with leprosy and the other derived from a patient with SLE (26). In another report it was found that the nucleotide sequence of an anti-Sm antibody derived from a patient with SLE (92) is identical to that of a cDNA clone from a fetal-liver cDNA library. This supports the view that genes encoding autoantibodies are also used by antibodies of different specificities and are most probably germ line encoded.

To recapitulate, V_H genes used by autoantibodies in both mice and humans (i) are often germ line encoded, (ii) have little polymorphism within the general population (possibly the result of evolutionary pressures) (122), and (iii) are expressed early in the development of the B-cell repertoire. A large fraction of self-reactive antibodies derive from structurally identical genes in normal individuals and in patients with autoimmune disease.

STRUCTURE OF IDIOTYPES ON AUTOANTIBODIES

Idiotypic determinants are defined serologically and can be located in the heavy chain, the light chain, or both chains. They can lie outside, close to, or within the antigen-combining site (paratope) of the antibody molecule. Therefore, since the interaction between idiotype and anti-idiotype may occur at different sites, in vitro assays to detect the idiotype directly depend upon their type of steric interactions. A precise structural definition of idiotopes is not only the prerequisite for a better classification and grouping of antibody molecules with similar antigen-binding specificity, but also necessary for the use of idiotypy as the target for manipulation of the immune system and specific therapy.

In this section we will review the idiotypic characterization of two types of autoantibody, human RFs and murine antithyroglobulin antibodies. This selection will allow us to illustrate the different methodological approaches used to address this issue.

Rheumatoid Factors

RFs, or antigammaglobulins, are heterogeneous populations of 19S IgM antibodies that react with epitopes expressed in the Fc portion of IgG. RFs are the most frequently detected human autoantibodies. Their reactivity with native monomeric IgG is usually of low affinity (ca. 10^4 to 10^5 mol/liter) (113). Although higher levels of RF activity are observed in patients with rheumatoid arthritis (RA), RFs do appear during many other diseases (syphilis, leprosy, liver cirrhosis, acquired immunodeficiency syndrome, etc.), i.e., chronic or acute inflammatory changes accompanied by the presence of circulating immune complexes (130).

Much of what we currently know about the molecular and genetic aspects of RFs is due to the pioneer work of Kunkel et al. (61, 63) and to the fact that ca. 5% of monoclonal IgM (paraproteins) in patients with Waldenström's macroglobulinemia have RF activity (99). In 1973 (61) it was shown that IgM RF Waldenström paraproteins from genetically unrelated individuals express not only individual idiotypes but also CRI. By using polyclonal antisera, two major CRI groups were identified, Wa and Po

Table 2. Positivity of Anti-Idiotype and Anti-Synthetic Peptide Antibodies in Human RFs

| Idiotype/ marker | Reactivity dependence | | | Disease association | | | |
	H	L	HL	Waldenström's macroglobulinemia	RA	Sjögren's syndrome	RF⁺ normal subjects
Wa	+	+	+	+	+		+
Po	+	+	+	+	+		+
Bla	+	+	+	+	+		+
17.109		+	+	+	+	+	±
6B6.6		+	+	+	+		±
Glo 86.3			+	+	+		−
G6	+		+	+	+		+
G8		+	+	+	+		+
PSL2		+		+	+	+	+
PSL3		+		+	±	+	±

(Table 2). These represent about 60 and 20% of IgM RF paraproteins, respectively. Recently, a third CRI group, Bla, was described (1). This is a minor idiotypic group that, unlike Wa and Po RFs, also reacts with DNA-histone. All three major CRI can be detected on polyclonal RFs in sera of RA patients and normal individuals (1, 35).

The original reports on the complete amino acid sequence of two members of the Wa group, SIE and WOL (4), and two members of the Po group, POM and LAY (14, 15), paved the way for the current understanding of the structural correlate for CRI expression. A marked homology among V_κIIIb light chains of Wa paraproteins and the V_HIII heavy chains of Po paraproteins suggested that the expression of these CRI depends on the presence of these light and heavy chains; i.e., CRI expression could correlate with a common structural feature within each CRI family. Although only one-third of Wa CRI IgM paraproteins display antigammaglobulin activity (45, 115), the V_κIII light chains are used by Wa CRI IgM RFs at much higher rate than in randomly collected myeloma immunoglobulins (11%) and normal human serum immunoglobulins (15%) (71, 106, 130).

Recently, Newkirk et al. (78–80) reported the complete amino acid sequence of two new members of the Wa CRI family, BOR and KAS, and two paraproteins, RIV (IgM) and SFL (IgG), which, on the basis of their primary structure, most probably belong to the Po CRI family (80, 81). According to these new studies it appears that members of the Wa⁺ CRI family share the V_κIIIb light chain and use different J_κ segments ($J_\kappa 1$, $J_\kappa 2$, and $J_\kappa 4$), related V_HI heavy chain, a D region of 9 or 10 amino acids, and the same J_H4 segment. Members of the Po family have not been analyzed as extensively. The four Po proteins sequenced (POM, LAY, RIV, and SFL) utilize a $V_H III$ subgroup gene segment. The amino acid sequences for POM and LAY are almost identical, whereas RIV differs in complementary determining region 2 (CDR2). POM, RIV, and SFL use related kappa light chains belonging to $V_\kappa IIIa$. LAY uses a light chain of the $V_\kappa I$ subgroup.

A new phase in the understanding of CRI of RFs occurred with the use of polyclonal antisera generated against synthetic peptides corresponding to complementary determining regions (CDRs) of the light and heavy chains of both the Wa and Po CRI. Antisera to CDR2 and CDR3 of the Wa SIE $V_\kappa IIIb$ light chains (PSL2 and PSL3) have been studied extensively. PSL2 was recognized in 20 of 25 RF paraproteins, and PSL3 was recognized in 15 (19, 20). Interestingly, anti-PSL2 antibodies also react with the Bla and some Po proteins that, as mentioned above, lack the Wa marker (2, 19, 20). These antisera were found to react with immunoglobulins devoid of RF activity, including a monoclonal antibody (EV1-15) specific for cytomegalovirus (80).

These results raise the question of whether antipeptide antibodies detect idiotypic determinants or linear epitopes encoded by germ line V_L or V_H genes. It is worth noting that chain recombination experiments in which $V_\kappa IIIb$ light chains of antibody EV1-15 (see above) were reassociated with the heavy chain of the Wa protein BOR indicated that the ensuing hybrid molecules acquired RF activity. Agnello and Barnes (2) showed that the Wa CRI is a conformational determinant whose expression depends on both heavy and light chains. Taken together, these results indicate that the specificity of Wa RFs depends in part on the association of a particular heavy chain with $V_\kappa IIIb$ light chains and that antipeptide antibodies appear to detect genetically related proteins by virtue of sequence homology.

Monoclonal antibodies constitute another stage in the analysis of the idiotopes expressed by human RFs. The first antibody used, 17.109 (16), was shown to react with about 50% of monoclonal RFs and only a small percentage of polyclonal normal immunoglobulins (34). Data collected from several different studies show that 17.109 is a conformational epitope expressed on intact IgM or isolated $V_\kappa IIIb$ light chains of Wa CRI proteins, possibly dependent on CDR1 or CDR3 residues close to the IgG-binding site. When antibody 17.109 was used to investigate the relatedness between monoclonal and polyclonal RFs from patients with RA, patients with Sjögren's syndrome, and healthy individuals, it was found that monoclonal RFs and polyclonal RFs from patients with Sjögren's syndrome express the 17.109 as well as the PSL2 and PSL3 epitopes, whereas RF-positive normal subjects express primarily the PSL2 epitope (19, 20, 34).

In another study (98), three monoclonal antibodies, 17.109, 6B6.6, and G6, were used in parallel to analyze 163 randomly collected IgM paraproteins of various specificities. Of these, 16% had RF activity. Among IgM, κ RFs 17.109 reacted in 27% of cases. The incidence of 17.109 CRI on non-RF IgM paraproteins was 11%. 6B6.6 monoclonal antibody reacted with 32% of randomly collected IgM RF paraproteins and with only 3% of non-RF paraproteins (24). High levels of 6B6.6 CRI were expressed in 33% of patients with RA. In no instance was this CRI paralleled by the expression of the 17.109 idiotype. This indicates that the 6B6.6 CRI is expressed on V_κIIIa light chains and that it probably represents a phenotypic marker of the V_κ328 germ line gene.

A monoclonal anti-idiotope, G6, was identified that reacts with 30% of IgM RFs, 9% of non-RF IgM paraproteins, and 5% of polyclonal IgM RFs (104). G6 is associated with various members of the Wa family and is frequently expressed on 17.109 CRI RFs. This idiotope seems to be dependent on conformation and is associated with the IgG-binding site. Since it is frequently associated with V_HI heavy chains, it is believed to be a phenotypic marker for this heavy chain gene subgroup. An interesting monoclonal anti-RF antibody is Glo 86.3. The idiotope that it defines is present on almost all Wa RF paraproteins and on polyclonal RFs in the sera of RA patients, with a high (ca. 50%) incidence of both IgM and IgG (67). Finally, monoclonal antibody G8 defines a conformation-dependent idiotope apparently linked to the presence of amino acids 94 to 100 of V_κIIIb light chains. This idiotope is variably expressed among highly related Wa IgM RFs and some normal immunoglobulins as well. The cumulative data that have been briefly discussed are summarized in Table 2.

Murine Antithyroglobulin Antibodies

To map idiotypy associated with autoantibodies, our group took a different approach. As a model system we used the idiotype 62 borne on murine autoantibodies to thyroglobulin. This is a recurrent, nondominant component (<2%) of the autoantibody response to thyroglobulin in BALB/c mice and is an autoimmunogen for B and T cells (124). Idiotype 62 thyroglobulin-binding antibodies can be found in the spleen cells of unmanipulated neonatal BALB/c mice (40). The interest in this system originates from the observation that idiotype 62 apparently distributes independently on the H and L chains (125, 127), and functions as a regulatory idiotype (85, 120) (see below).

The independent expression of this idiotype on the H chain was confirmed by transfecting the cloned V_H62 rearrangement into J558L myeloma cells and demonstrating that the expressed immunoglobulin (cy_162) expresses idiotype 62 (105). Taking advantage of this fact, we decided to use site-directed mutagenesis to analyze idiotype expression. The rationale for

Table 3. Comparison of the Primary Structures of the V_H of Idiotype 62-Positive and Idiotype 62-Negative Monoclonal Antibodies

```
                                               CDR1
                                               30
62       DVKLVESGGGLVKLGGSLKLSCAASGFTFSRYYMSWVRQTPEKRLELVAA
A26      ------------------------------S------------------
B13      ------------------------------S------------------
MOPC21   --Q---------QPRE--------------SFG--H----A---G---W-Y
66E-3    E-M------V-MEP-----------------A--------------W-T

         CDR2
                      60                                    90
62       INSNGGSTYYPDTVKGRFTISRDNAKNTLYLQMSSLKSEDTALYYCAR
A26      ------------------------------------------------
B13      ------------------------------------------------
MOPC21   -S-GSSTLH-A-------------P----F---T--R----M-----
66E-3    -S-G-S-HLPSRQCE---------------------R----M-----

         CDR3                                           Id
62       KAYSHG      MDYWGQGTS                           +
A26      HELTGTALFA---------LV                           -
B13      QALLSIHYFA---------TL                           -
MOPC21   WGNPYYA     ---------                           -
66E-3    PPLSLLDAYA---------                             -
```

designing the mutations was provided by the comparative analysis of several antibodies whose V_H share identical framework residues with V_H62 but lack idiotype 62 expression. These show marked differences clustered in the CDR (Table 3). Mutated V_H genes were subcloned into an appropriate expression vector and electroporated into the J558L myeloma cells. The purified genetically engineered immunoglobulins were then subjected to analysis for idiotype 62 expression by enzyme-linked immunosorbent assay inhibition (104a).

The results are summarized in Table 4 and indicate that idiotype 62 expression was abrogated in CDR3 mutants. Both CDR3 mutants 94VP, in which residue Ala-94 was replaced with a Val with the insertion of a Pro, and 95N, in which Tyr-95 was replaced with an Asn residue, lost idiotypic expression. The involvement of CDR3 is confirmed by the fact that $\gamma1$ NANP, a mutant in which 12 amino acid residues were inserted at position 95, also lost idiotype expression. The CDR1 deletion mutant ($\Delta1$) in which amino acid residues Tyr-33 and Met-34 were deleted no longer expresses idiotype 62. It is likely that deletion of two amino acids in CDR1 determines a conformational change in the V_H fold. Similarly, idiotype expression was abrogated by substituting Met-34 with Lys. Met-34 is indeed an important residue for the packing of the V_H domain. By contrast, when Tyr-33 was

Table 4. Idiotype 62 Expression of Recombinant V_H62 Mutant Antibodies

Transfectoma	Site mutation	Type of amino acid mutation	Change	Position	Idiotype 62
γ1 wt	None	None	None		+
γ1 Δ133/34	CDR1	Deletion	2 amino acids	33–34	−
γ1 33S	CDR1	Substitution	Y → S	33	+
γ1 34K	CDR1	Substitution	M → K	34	−
γ1 52bI	CDR2	Substitution	N → I	52b	+
γ1 56D	CDR2	Substitution	Y → D	56	+
γ1 94VP	CDR3	Substitution/insertion	A → V/P	94/95	−
γ1 95N	CDR3	Substitution	Y → N	95	−
γ1 NANP	CDR3	Insertion	(NANP)₃	95	−

replaced by a Ser residue, there was no effect on idiotype 62 expression. Construction and analysis of single-amino-acid substitutions in CDR2, namely, 52b-Asn → Ile and 56-Tyr → Asp, revealed no effect on idiotype 62 expression. Thus, idiotype 62 expression appears to be dependent mainly upon amino acid residues in CDR3.

General Considerations

From the above examples it is clear that the process of defining idiotypy of autoantibodies in molecular terms is difficult and serendipitous. Although the advent of new technologies has considerably accelerated our analytical ability, definitive answers are still missing. In most instances these will also require information at the three-dimensional level that can be obtained only by using crystallography and nuclear magnetic resonance techniques.

FUNCTIONAL CHARACTERISTICS OF THE AUTOIMMUNE NETWORK

The maintenance of tolerance to self is likely to depend upon a delicate equilibrium among various elements of the internal environment, namely, autoantigens, idiotypes, and anti-idiotypes. This finely tuned homeostatic mechanism can be perturbed by (i) changes in the relative concentration of any of these three elements and (ii) the interaction between antibodies and receptors on cells of either the immune system or other systems in the organism. We refer to this as the autoimmune network (120), i.e., the comprehensive framework for the body of interactions among structures of the internal milieu based on chemical complementarity. Balance within such a self-centered immune system is, therefore, a highly complex phenomenon that constantly oscillates between physiology and pathology. Figure 2 schematically illustrates the spectrum of possibilities involving the idiotype–anti-idiotype interactions. At one end are the aspects of a merely regulatory nature; at the other are conditions that more directly pertain to pathology. In this section, we will discuss some of the experimental observations which have inspired this view. We will examine first the interactions that pertain to the possible involvement of anti-idiotypes and idiotypes in the activation of the autoimmune network.

Induction of Autoimmunity by Network Manipulation

Induction by anti-idiotypes

The possibility of perturbing the normal, tolerant state of an animal by using anti-idiotype as a route to inducing de novo autoantibodies of predetermined specificity has been explored in various laboratories. Possibly, common to all these experiments was the idea that antibodies to the idiotype of autoantibodies could substitute for the autoantigen and activate self-reactive clones.

Several years ago we reported (128) on the immunization of naïve adult BALB/c mice with purified rabbit antibodies to idiotype 62, the idiotype of a minor fraction of autoantibodies to thyroglobulin in this strain. We found that the majority of the anti-idiotype-immunized mice produced significant levels of idiotype and about half of them also produced autoantibodies to thyroglobulin. The same anti-idiotypic antibodies were subsequently used to immunize animals other than mice. Initial studies used young thyroiditis-free Buffalo rats. Here again, all anti-idiotype 62-immunized rats produced idiotype 62/thyroglobulin-binding autoantibodies. Comparable results were subsequently obtained in the Fischer rat strain, which, unlike the Buffalo strain, is not an autoimmune-prone strain. Interestingly, the perturbation induced by anti-idiotype immunization remained detectable several months after the initial immunization.

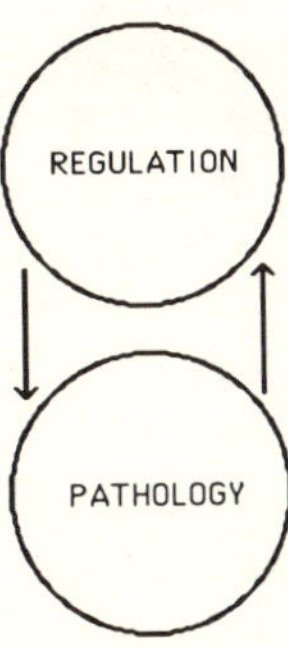

Figure 2. Functions of the idiotype–anti-idiotype interactions.

Of a similar nature are the results of experiments by Teitelbaum et al. (109) with SLE-prone mice. They injected MRL/$^{++}$ mice intravenously with F(ab')$_2$ fragments of rabbit antibodies to a monoclonal autoantibody to DNA bearing a CRI for anti-DNA antibodies. This treatment resulted in an increased level of CRI in serum and a small but significant increase in the level of circulating anti-DNA autoantibodies. CRI and anti-DNA antibodies in anti-idiotype-injected mice persisted for 4 weeks. In another experiment it was shown that normal BALB/c mice injected with anti-anti-DNA antibodies began to express the idiotype of anti-DNA antibodies in their serum.

The anti-idiotypic approach has also been used to induce polyreactive autoantibodies (31). Thus, a rabbit immunized with a murine anti-idiotype antibody directed against a polyendocrine-reactive human autoantibody produced autoantibodies with a reactivity similar to that of the initial human autoantibody.

A pathogenic potential of anti-idiotypic antibodies to anti-DNA autoantibodies was shown more recently by Mendlovic et al. (73). They immunized C3H.SW female mice with a murine monoclonal anti-idiotype identifying the 16/6 idiotype of human anti-DNA autoantibodies. Both strains developed an SLE-like disease characterized by the appearance of wide-spectrum antinuclear antibodies and high levels of 16/6 idiotype. Similarly, immunization with anti-idiotype antibodies recognizing major CRI (H130) on anti-DNA antibodies from MRL-lpr/lpr strain induced H130 expression and a rise in serum anti-DNA antibodies in both MRL/$^{++}$ and BALB/c mice, although only 10% of immunoglobulins expressing H130 showed anti-DNA activity (109).

Induction by idiotype

The report by Hall et al. (44) on the passive transfer of interstitial nephritis with isologous autoantibodies to tubular basement membrane (TBM) antigens can be viewed as the first example of induction of autoantibodies by the idiotype itself. In this experiment normal guinea pigs given isologous IgG1 and IgG2 anti-TBM autoantibodies and tested 14 days after the passive transfer showed a high titer of anti-TBM autoantibodies. Since these were of a different isotype from those injected, a plausible explanation would be that passively transferred autoantibodies endogenously stimulated (via their idiotypes) the production of anti-TBM autoantibodies.

In recent years this approach has been used to induce anti-DNA autoantibodies and clinical manifestations of lupus. In these experiments normal adult C3H.SW (72) and BALB/c (8) mice were immunized with the major 16/6 CRI of human anti-DNA antibodies. In both instances mice developed an SLE-like disease characterized by the appearance of high titers of anti-DNA and other autoantibodies (Sm/RNP, SS-A, SS-B, and histone), mild leukopenia, and lupus nephritis with deposits of immunoglobulins along the glomerular basal membrane. Animals also produced high titers of idiotype 16/6 antibodies, presumably induced via a perturbation of the autoimmune network similar to that observed in the experiment by Hall. In (NZB × NZW)F$_1$ mice, immunization with the 16/6 idiotype caused an acceleration of the disease. Notably, the effect was restricted to the 16/6 idiotype, since immunization of mice with another idiotype (PME 77), expressed only on a minor proportion of anti-DNA antibodies, failed to influence the production of anti-DNA antibodies (55).

Autoantibodies as Anti-Idiotypes to Exogenous Pathogens

Several years ago we speculated (120) that a specific perturbation of the autoimmune network could originate from an initial generation of antibodies directed against exogenous unrelated antigens (e.g., bacteria, viruses, and drugs) which fortuitously share idiotypy with paratopes for autoantigens. In turn, the anti-idiotypes to these molecules could presumably stimulate complementary receptors irrespective of their antigen specificity. The importance of such a pathway in perturbing the autoimmune network and establishing a state of autoreactivity is self-evident. Experimental evidence was subsequently provided by Dwyer et al. (27), who demonstrated that anti-idiotypes of the antidextran immune response are autoantibodies to the ACh receptor. Since dextran is the antigenic determinant of *Enterobacter cloacae,* antireceptor antibodies were generated as a part of an anti-idiotypic response to exogenous pathogens.

Anti-Idiotypes as Antireceptor Autoantibodies

Evidence now exists that anti-idiotype antibodies can function as antireceptor antibodies on cells other than lymphoid cells. The first demonstration was made by Sege and Peterson (98), who showed that rabbit anti-idiotypes to rat antibodies against bovine insulin behave like insulin. Competition experiments with insulin and anti-idiotype antibodies for binding to insulin receptors on rat epididymal fat cells showed that the antibodies were able to cause almost complete inhibition of insulin binding to the receptor. Subsequently, Shechter et al. (100) showed that immunization of mice with insulin itself may be sufficient to stimulate the production of both anti-insulin and anti-insulin receptor (anti-anti-insulin) antibodies. The latter were characterized as anti-idiotypic antibodies of the mirror image type, since they (i) displaced labeled insulin from specifically binding to fat cells and (ii) mimicked the biological effect of insulin at the level of glucose oxidation and lipolysis. How often this mechanism may apply to the generation of autoantibodies to receptors in humans is not known, although a similar mechanism has been considered to be involved in patients with the rare association of type I insulin-dependent diabetes with *Acantosis nigricans.*

Using a similar approach, Wasserman et al. (114) induced in rabbits symptoms equivalent to myasthenia gravis in humans. A potent agonist of the ACh receptor, a derivative of BisQ [*trans*-3,3′-bis-α-(trimethylam-monio)methylazobenzene bromide], was used as the immunogen. Rabbits immunized with purified anti-BisQ antibodies yielded antisera which recognized rat, torpedo, or eel ACh receptor. This binding was inhibited by the free ligand BisQ. Some rabbits showed signs of muscle weakness similar to that currently seen after immunization with ACh receptor, and the intra-muscular injection of neostigmine alleviated the symptoms of disease. Similarly, mice immunized with BisQ anti-idiotype antibodies developed auto-antibodies with anti-ACh receptor activity (23).

Analogous demonstrations have been made in two other systems, the thyroid-stimulating hormone (TSH) receptor (6, 53) and the antiadrenergic receptor. In the first case, rabbit anti-idiotype antibodies to rat or human antibodies to the TSH receptor of Graves' disease patients inhibited the binding of the idiotypic antibodies to the TSH receptor, the binding of TSH to thyroid plasma membranes, and the in vitro TSH-mediated adenylate cyclase stimulation. In the second case (96), anti-idiotypes to the antiadren-ergic receptor bound the antiadrenergic receptor of various cell types, inhibited the binding of alprenolol to the membrane-bound antiadrenergic receptor in a dose-dependent manner, and modulated catecholamine-sensitive adenylate cyclase.

Idiotype–Anti-Idiotype Immune Complexes

Naturally occurring idiotype–anti-idiotype complexes were first found by Morgan et al. (74) in the sera of 19 of 23 healthy subjects. A subsequent study by Geltner et al. (38) detected an anti-idiotypic activity in the IgM fraction of mixed cryoglobulins from 11 patients with mixed cryoglobulin-emia. The natural formation of circulating idiotype–anti-idiotype during an immune response was found by Rose and Lambert (91) in BALB/c mice immunized with pneumococcus vaccine, which bears phosphocholine as a major antigenic determinant. BALB/c mice respond to this vaccine by producing antiphosphocholine antibodies which express a typical cross-reacting idiotype (T15). An analysis of the dynamics of idiotype–anti-idio-type complex formation indicated that they first form in a relative excess of idiotype and, subsequently, concomitant with the increased production of anti-idiotype antibodies, in an excess of anti-idiotype.

Several experiments have also been done to investigate the potential participation of idiotype–anti-idiotype complexes in the immunopathologic consequences of immune complex diseases. Goldman et al. (41) studied the generation of idiotype–anti-idiotype immune complexes in the kidneys during the antiphosphocholine response elicited by polyclonal B-cell activation by a lipopolysaccharide. By the immunofluorescence technique, both idiotype-bearing immunoglobulins and anti-idiotype-bearing immunoglobulins were detected in the glomerular deposits of lipopolysaccharide-treated mice. Elution studies further demonstrated the existence of idiotype and anti-idiotype in the nephritic kidneys. The presence of anti-idiotype antibodies within tissue-bound immune complexes was sought by Zanetti and Wilson (129) in a rabbit with chronic serum sickness glomerulonephritis induced by daily intravenous injection of bovine serum albumin, a model that constitutes the prototype for immune complex-mediated glomerulonephritis. The 7S IgG fraction of the kidney eluate was demonstrated to have auto-anti-idiotypic activity. However, it was not determined whether idiotype–anti-idiotype complexes formed in the circulation or at the level of the tissue, since the dynamics of immune-complex formation and tissue deposition in this model are subject to continuous variation dependent upon the relative concentration of each component in serum.

Suppression of Autoimmunity by Anti-Idiotype and Idiotype

For practical purposes, possibly the most important aspect in studying the idiotype of autoantibodies is their use in specifically down regulating autoreactive clones. To this end, several reports have appeared in which experimentally induced or spontaneously occurring autoimmune diseases have been down regulated by manipulation of the idiotypic network. These approaches can be separated into two categories depending on whether anti-idiotype or idiotype itself is used.

Suppression by administration of anti-idiotype

The first report on spontaneously occurring autoimmune diseases was by Zanetti and Bigazzi (118), who studied autoimmune thyroiditis of BUF rats. Using rabbit antibodies specific for the CRI (CRI ART) of serum autoantibodies to thyroglobulin, they showed that an anti-idiotype treatment preceded by sublethal X-irradiation of the animals could suppress the ongoing production of autoantibodies over a period of 7 weeks.

Hahn and Ebling (42) subsequently showed that in the autoimmune-prone lupus mice, the spontaneous production of antibodies to double-stranded DNA in (NZB/NZW)F$_1$ mice can also be down regulated by passive anti-idiotypic immunity. Treatment of 4- to 6-week-old mice by repeated intraperitoneal injections of a monoclonal anti-idiotype antibody directed against a CRI of anti-DNA antibodies resulted in specific suppression of antibodies to double-stranded DNA but not single-stranded DNA.

This treatment partially suppressed the proliferative nephritis typical of these mice and increased their survival rate.

The first studies of experimentally induced autoimmunity were done by using as a model interstitial nephritis in rodents, in which autoantibodies to antigens of the kidney TBM are considered to be of causative importance. Brown et al. (13) first reported partial suppression of the anti-TBM response in guinea pigs treated with rabbit anti-idiotypic antibodies specific for anti-TBM autoantibodies. At the same time, a decrease in the immunopathologic lesions was also observed. Subsequent experiments with rats (126) by a somewhat similar approach further analyzed the specificity of the effect on autoantibody production. In this disease, antibodies reactive with autologous TBM antigens react with a 42,000-dalton TBM glycoprotein recovered by collagenase solubilization of Brown Norway rat TBM. Rabbit anti-idiotypes prepared against anti-TBM antibodies eluted from diseased kidneys and injected intraperitoneally 48 h prior to induction of the renal disease selectively suppressed the production of autoantibodies to TBM. Notably, they principally suppressed autoantibodies to autologous noncollagenous TBM glycoprotein moieties. In treated rats the lymphomonocytic infiltration of the cortical interstitium, typical of this autoimmune disease, was decreased in parallel with the degree of suppression of autoantibody production.

Suppression by administration of idiotype

The possibility of inducing specific unresponsiveness via autologous idiotype immunization in various antigen systems was demonstrated in several laboratories. Soluble idiotype, cell surface-bound idiotype, and idiotype in adjuvant (vaccine) have all been used.

Neilson and Phillips (77), studying interstitial nephritis in rats, first showed that a state of active anti-idiotypic immunity against autoantibodies to TBM antigens could be generated by injecting normal rats with autoantigen-reactive (idiotype-positive) T lymphoblasts emulsified in adjuvant. These rats were partially protected from disease induced by a subsequent immunization with TBM. Circulating autoantibodies and immunopathologic lesions were suppressed over controls. Protected animals showed a decreased T-cell proliferative response to TBM, produced autoantibodies with reduced binding affinity, and mounted a specific anti-idiotypic response that was probably responsible for the marked attenuation of the disease. Subsequently, Neilson et al. showed (76) that anti-idiotype-mediated suppression could be transferred by cells if the donor animals had been pretreated with low-dose cyclophosphamide, a drug known to abrogate suppressor T cells. Interestingly, anti-idiotypic immunity could not be detected in rats with interstitial nephritis except in those given low-dose cyclo-phosphamide treatment.

Experiments have also been done with the SLE-prone (NZB × NZW) F_1 mice (42). Repeated injections of a monoclonal antibody to double-stranded DNA caused specific suppression of autoantibodies to double- but not single-stranded DNA. The suppressive effect was associated with the appearance of antibodies to the idiotype of the immunizing anti-DNA monoclonal antibody. Anti-idiotype-producing mice were partially protected from proliferative nephritis.

Recently, De Kozak et al. (25) showed the effectiveness of idiotype immunization in a rat model of uveoretinitis. Preimmunization of Lewis rats with monoclonal antibodies to the retinal autoantigen (S antigen) resulted in production of high-titer circulating auto-anti-idiotypes and long-lasting protection from inflammatory changes of the retina.

The relative importance of cellular and humoral anti-idiotypic immunity, respectively, in the suppression of autoreactivity was analyzed by our group (124) by using BALB/c mice actively immunized with idiotype 62. Briefly, after the induction of high-titer circulating anti-idiotype 62 antibodies, mice were challenged with a low dose of antigen (thyroglobulin) in adjuvant. By measuring the ensuing autoantibody response, we found that idiotype 62-immunized mice produced markedly lower levels of autoantibodies than normal mice or mice immunized with an irrelevant idiotype. To rule out the interference by circulating idiotype 62, we performed adoptive transfers. Transfer of separate T- and B-cell populations revealed that idiotype-primed B lymphocytes could transfer the ability to make auto-anti-idiotypes and idiotype-positive molecules but not to suppress the autoantibody response to thyroglobulin. On the other hand, idiotype 62-primed T lymphocytes were sufficient to transfer suppression of the autoantibody response to thyroglobulin. Taken together, these results suggest that autologous idiotype activates both the B- and T-cell compartments and that specific suppression by idiotype autoimmunization best correlates with the activation of idiotype-responsive T cells.

CONCLUSIONS

The discovery of idiotypy 25 years ago has opened a new dimension in the analysis of autoimmunity. The topics reviewed herein are eloquent proof of the considerable effort made by many investigators to understand the relation between idiotypy and self reactivity and to elucidate the role of idiotypy in regulation of autoimmunity. Abundant evidence was provided to indicate that idiotypes of self-reactive V regions may be involved in the regulation of ongoing or induced responses to self antigens. The studies that were discussed, however, presented only one aspect at the time when defined experimental conditions were used. Therefore, the way in which the autoimmune network functions as a whole and the prevalent forces that regulate it at the physiological level are still unclear. We have stressed the fact that the internal environment balance between physiology and pathology may be a delicate one. This equilibrium may be perturbed by a variety of factors that are still unknown, although compelling evidence indicates that the autoimmune network is endowed with the ability to efficiently regulate itself. Although a precise understanding of the physiological role of idiotype network interactions in the regulation of autoimmunity requires additional studies, intervention and therapy with idiotypy may become a reality in the near future, provided that progress is made concerning our ability to resolve the structure of idiotypy and understand how idiotypes are handled at the B- and T-cell level during the immune response.

Acknowledgments. This work was supported by Public Health Service grant AL 23871 from the National Institutes of Health. M.Z. is a scholar of the Leukemia Society of America, Inc. N.D. was supported by a grant from the Scientific Foundation of the republic of Serbia, Yugoslavia. P.L. is recipient of a fellowship from the International Research & Exchange Board.

REFERENCES

1. **Agnello, V., A. Arbetter, G. Ibanez De Kasep, R. Powell, E. M. Tan, and F. Joslin.** 1980. Evidence for a subset of rheumatoid factors that cross-react with DNA-histine and have a distinct cross-idiotype. *J. Exp. Med.* **151**:1514–1527.

2. **Agnello, V., and J. L. Barnes.** 1986. Human rheumatoid factor cross-idiotypes. I. Wa and Bla are heat labile conformational antigens requiring both heavy and light chains. *J. Exp. Med.* **164**:1809–1814.

3. **Andrejewski, C., Jr., J. Rauch, E. Lafer, B. D. Stollar, and R. S. Schwartz.** 1981. Antigen-binding diversity and idiotypic cross-reactions among hybridoma autoantibodies to DNA. *J. Immunol.* **126**:226–231.

4. **Andrews, D. W., and J. D. Capra.** 1981. Complete amino acid sequence of variable domains from two monoclonal human anti-gamma globulins of the Wa cross-idiotypic group: suggestion that the J segments are involved in the structural correlate of the idiotype. *Proc. Natl. Acad. Sci. USA* **78**:3799–3803.

5. **Baird, L. G., T. G. Tachovsky, M. Sandberg-Wollheim, H. Koprowski, and A. Nisonoff.** 1980. Identification of a unique idiotype in cerebrospinal fluid and serum of a patient with multiple sclerosis. *J. Immunol.* **124**:2324–2328.

6. **Baker, J. R., Y. G. Lukes, and K. D. Burman.** 1984. Production, isolation and characterization of rabbit anti-idiotypic antibodies directed against human antithyrotropin receptor antibodies. *J. Clin. Invest.* **74**:488–495.

7. **Berman, J. E., S. J. Mellis, R. Pollock, C. Smith, H. Y. Suh, B. Heinke, C. Kowal, U. Surti, L. Chess, C. Cantor, and F. W. Alt.** 1988. Content and organization of the human Ig V_H locus: definition of three new V_H families and linkage to the Ig CH locus. *EMBO J.* **7**:727–737.

8. **Blank, M., S. Mendlovic, E. Mozes, and Y. Shoenfeld.** 1988. Induction of SLE-like disease in naive mice with a monoclonal anti-DNA antibody derived from a patient with polymyositis carrying the 16/6 ID. *J. Autoimmun.* **1**:683–691.

9. **Bona, C. A.** 1987. p. 152–168. *In* C. A. Bona (ed.), *Regulatory Idiotypes.* John Wiley & Sons, Inc., New York.

10. **Bona, C. A.** 1988. V genes encoding autoantibodies: molecular and phenotypic characteristics. *Annu. Rev. Immunol.* **6:**327–358.

11. **Bona, C. A., S. Finley, S. Waters, and H. G. Kunkel.** 1982. Anti-immunoglobulin antibodies. III. Properties of sequential anti-idiotypic antibodies to heterologous anti-γ globulins. Detection of reactivity of anti-idiotype antibodies with epitopes of Fc fragments (homobodies) and with epitopes and idiotopes (epibodies). *J. Exp. Med.* **156:**986–999.

12. **Brodeur, P., G. E. Osman, J. J. Makle, and T. M. Lalor.** 1988. The organization of the mouse IgH-V locus. *J. Exp. Med.* **168:**2261–2278.

13. **Brown, C. A., K. Carey, and R. B. Colvin.** 1979. Inhibition of auto-immune tubulointerstitial nephritis in guinea pigs by heterologous antisera containing anti-idiotype antibodies. *J. Immunol.* **123:**2102–2107.

14. **Capra, J. D., and J. M. Kehoe.** 1974. Structure of antibodies with shared idiotype: the complete sequence of the heavy chain variable regions at two immunoglobulin and anti-gammaglobulins. *Proc. Natl. Acad. Sci. USA* **71:**4032–4036.

15. **Capra, J. D., and D. J. Klapper.** 1976. Complete amino acid sequence of the variable domains of two human IgM anti-gammaglobulins (LAY/POM) with shared idiotypic specificities. *Scand. J. Immunol.* **5:**677–684.

16. **Carson, D. A., and S. Fong.** 1983. A common idiotype on human rheumatoid factors identified by a hybridoma antibody. *Mol. Immunol.* **20:**1081–1087.

17. **Caulfield, M. J., D. Stanko, and C. Calkins.** 1989. Characterization of the spontaneous autoimmune (anti-erythrocyte) response in NZB mice using a pathogenic monoclonal autoantibody and its anti-idiotype. *Immunology* **66:**233–237.

18. **Chen, P. P., K. Albrandt, T. J. Kipps, V. Radoux, F.-T. Liu, and D. A. Carson.** 1987. Isolation and characterization of human VKIII germ-line genes. Implications for the molecular basis of human VKIII light chain diversity. *J. Immunol.* **139:**1727–1733.

19. **Chen, P. P., J. H. Fong, F. Goni, R. A. Houghten, B. Frangione, F-T. Liu, and D. A. Carson.** 1987. Analysis of human rheumatoid factors with anti-idiotypes induced by synthetic peptide. *Monogr. Allergy* **22:**12–23.

20. **Chen, P. P., F. Goni, S. H. Fong, F. Jirik, J. H. Vaughan, B. Frangione, and D. A. Carson.** 1985. The majority of human monoclonal IgM rheumatoid factors express a "primary structure-dependent" cross-reactive idiotype. *J. Immunol.* **134:**3281–3285.

21. **Chen, P. P., R. A. Houghten, S. Fong, G. H. Rhodes, T. A. Gilbertson, J. H. Vaughan, R. A. Lerner, and D. A. Carson.** 1984. Anti-hypervariable region antibody induced by a defined peptide: an approach for studying the structural correlates of idiotypes. *Proc. Natl. Acad. Sci. USA* **81:**1784–1788.

22. **Chen, P. P., D. L. Robbins, F. R. Jirik, T. J. Kipps, and T. J. Carson.** 1987. Isolation and characterization of a light chain variable region gene for human rheumatoid factors. *J. Exp. Med.* **166:**1900–1905.

23. **Cleveland, W. L., N. H. Wasserman, R. Sarangarajan, A. S. Penn, and B. F. Erlanger.** 1983. Monoclonal antibodies to the acetylcholine receptor by a normally functioning auto-anti-idiotypic mechanism. *Nature* (London) **305:**56–57.

24. **Crowley, J. J., R. D. Goldfien, R. E. Schrohenloher, H. L. Spiegelberg, G. J. Silverman, R. A. Mageed, R. Jefferis, W. J. Koopman, D. A. Carson, and S. Fong.** 1988. Incidence of three cross-reactive idiotypes on human rheumatoid factor paraproteins. *J. Immunol.* **140:**3411–3418.

25. **De Kozak, Y., M. Mirshahi, C. Boucheix, and J.-P. Faure.** 1987. Prevention of experimental autoimmune uveoretinitis by active immunization with autoantigen-specific monoclonal antibodies. *Eur. J. Immunol.* **17:**541–554.

26. **Dersimonian, H., R. S. Schwartz, K. J. Barrett, and D. B. Stollar.** 1987. Relationship of human variable region heavy chain germline genes to genes encoding anti-DNA autoantibodies. *J. Immunol.* **139:**2496–2501.

27. **Dwyer, D. S., M. Vakil, and J. F. Kearney.** 1986. Idiotypic network connectivity and a possible cause of myasthenia gravis. *J. Exp. Med.* **164:**1310–1318.

28. **Ebers, G. C., J. B. Zabriskie, and H. G. Kunkel.** 1979. Oligoclonal immunoglobulins in subacute sclerosing panencephalitis and multiple sclerosis: a study of idiotypic determinants. *Clin. Exp. Immunol.* **35:**67–75.

29. **Ehrlich, P., and J. Morgenroth.** 1900. *The Collected Papers of Paul Ehrlich—1900,* p. 205–255. F. Himmelweit (ed.). Pergamon Press, Oxford. (Republished in 1959.)

30. **Elrehewy, M., Y. M. Kong, A. A. Giraldo, and N. R. Rose.** 1981. Syngeneic thyroglobulin is immunogenic in good responder mice. *Eur. J. Immunol.* **11:**146–151.

31. **Essani, K., J. Srinivasappa, R. McClintok, B. S. Prabhakar, and A. L. Notkins.** 1986. Multi-organ reactive IgG antibody induced by anti-idiotypic antibody to a human monoclonal IgM autoantibody. *J. Exp. Med.* **163:**1355–1360.

32. **Evans, S. W., T. Feizi, R. Childs, and N. R. Ling.** 1983. Monoclonal antibody against a cross-reactive idiotypic determinant found on human autoantibodies with anti-I and -i specificities. *Mol. Immunol.* **20:**1127–1131.

33. **Evans, S. W., N. A. Gregson, S. L. Leibowitz, and N. R. Ling.** 1984. Investigation of the idiotypic determinants of IgM monoclonal proteins associated with neuropathy. *Clin. Exp. Immunol.* **57:**621–625.

34. **Fong, S., P. P. Chen, T. A. Gilbertson, J. R. Weber, R. I. Fox, and D. A. Carson.** 1986. Expression of three cross-reactive idiotypes on rheumatoid factor antibodies from patients with autoimmune diseases and seropositive adults. *J. Immunol.* **137:**122–128.

35. **Forre, O., J. H. Dobloug, T. E. Michaelsen, and J. B. Natvig.** 1979. Evidence of similar idiotypic determinants on different rheumatoid factor populations. *Scand. J. Immunol.* **9:**281–289.

36. **Fritz, R. B., and A. E. Desjardins.** 1982. Idiotypes of Lewis rat antibodies to encephalitogenic peptides of guinea pig myelin basic protein: in vitro and in vivo studies. *J. Immunol.* **128:**247–250.

37. **Gaither, K. K., and J. B. Harley.** 1989. A shared idiotype among human anti-Ro/SSA autoantibodies. *J. Exp. Med.* **169:**1583–1588.

38. **Geltner, D., E. C. Franklin, and B. Frangione.** 1980. Anti-idiotypic activity in the IgM fractions of mixed cryoglobulins. *J. Immunol.* **125:**1530–1535.

39. **Glotz, D., M. Sollazzo, S. Riley, and M. Zanetti.** 1988. Isotype, V_H genes, and antigen-binding analysis of hybridomas from newborn normal BALB/c mice. *J. Immunol.* **141:**383–390.

40. **Glotz, D., and M. Zanetti.** 1986. Detection of a regulatory idiotype on a spontaneous neonatal self-reactive hybridoma antibody. *J. Immunol.* **137:**223–227.

41. **Goldman, M., L. M. Rose, A. Hochmann, and P. H. Lambert.** 1982. Deposition of idiotype-anti-idiotype immune complexes in renal glomeruli after polyclonal B cell activation. *J. Exp. Med.* **155:**1385–1399.

42. **Hahn, B. H., and F. M. Ebling.** 1983. Suppression of NZB/NZW murine nephritis by administration of a syngeneic monoclonal antibody to DNA. *J. Clin. Invest.* **71:**1728–1736.

43. **Hahn, B. H., and F. M. Ebling.** 1984. A public idiotypic determinant is present on spontaneous cationic IgG antibodies to DNA from mice of unrelated lupus-prone strains. *J. Immunol.* **133:**3015–3019.

44. **Hall, T. L., R. B. Colvin, K. Carey, and R. T. McCluskey.** 1977. Passive transfer to autoimmune disease with isologous IgG_1 and IgG_2 antibodies of the tubular basement membrane in strain XIII guinea pigs. *J. Exp. Med.* **146:**1246–1260.

45. **Hardy, R. R., C. E. Carmack, S. A. Shinton, R. Riblet, and K. A. Hayakawa.** 1989. Single V_H gene is used predominantly in anti-BrMRBC hybridomas derived from purified Ly-1 B cells. *J. Immunol.* **142:**3643–3651.

46. **Hildebrandt, S., E. S. Weiner, W. C. Earnshaw, M. Zanetti, and N. F. Rothfield.** 1990. Idiotypic analysis of human anti-centromere autoantibodies: further evidence for an autoantigen driven immune response. *Autoimmunity,* in press.

47. **Humphries, C., A. Shen, W. A. Kuziel, J. D. Capra, F. R. Blattner, and P. Tuker.** 1988. Characterization of a new human immunoglobulin V_H family that shows preferential rearrangement in immature B cell tumors. *Nature* (London) **331:**446–449.

48. **Hunkapiller, T., and L. Hood.** 1989. Diversity of the immunoglobulin gene superfamily. *Adv. Immunol.* **44:**1–62.

49. **Ilowite, N., J. F. Wedgewood, and V. Bonagura.** 1989. Expression of the major rheumatoid factor cross-reactive idiotype in juvenile rheumatoid arthritis. *Arthritis Rheum.* **32:**265–270.

50. **Isenberg, D., C. Dudeney, F. Wojnaruska, B. S. Bhogal, J. Rauch, A. Schattner, Y. Naparstek, and D. Duggan.** 1985. Detection of cross reactive anti-DNA antibody idiotypes on tissue-bound immunoglobulins from skin biopsies of lupus patients. *J. Immunol.* **135:**261–264.

51. **Isenberg, D. A., and C. Colins.** 1985. Detection of cross-reactive anti-DNA antibody idiotypes on renal tissue-bound immunoglobulins from lupus patients. *J. Clin. Invest.* **76:**287–294.

52. **Isenberg, D. A., Y. Shoenfeld, M. P. Madaio, J. Rauch, M. Reichlin, B. D. Stollar, and R. S. Schwartz.** 1984. Anti-DNA antibody idiotypes in SLE. *Lancet* **ii:**417–422.

53. **Islam, M. N., B. M. Pepper, R. Briones-Urbina, and N. R. Farid.** 1983. Biological activity of anti-thyrotropin anti-idiotypic antibody. *Eur. J. Immunol.* **13:**57–63.

54. **Izui, S., P. H. Lambert, G. Fournie, H. Turler, and P. A. Miescher.** 1977. Features of systemic lupus erythematosus in mice injected with bacterial lipopolysaccharides. Identification of circulating DNA and renal localization of DNA-anti-DNA complexes. *J. Exp. Med.* **145:**1115–1130.

55. **Jacob, L., and F. Tron.** 1984. Induction of anti-DNA autoanti-idiotypic antibodies in (NZB × NZW)F1 mice: possible role for specific immune suppression. *Clin. Exp. Immunol.* **58:**293–299.

56. **Jerne, N. K.** 1974. Towards a network theory of the immune system. *Ann. Immunol.* (Paris) **125:**373–389.

57. **Kalunian, K. C., N. Panosian-Sahakian, F. M. Ebling, A. H. Cohen, J. S. Louie, J. Kaine, and B. H. Hanh.** 1989. Idiotypic characteristics of immunoglobulins associated with systemic lupus erythematosus. *Arthritis Rheum.* **32:**513–522.

58. **Kasturi, K., M. Monastier, A. Mayer, and C. Bona.** 1988. Biased use of certain V_k gene families by autoantibodies and their polymorphism in autoimmune mice. *Mol. Immunol.* **25:**213–219.

59. **Kipps, T. J., S. Fong, E. Tomhave, P. P. Chen, R. D. Goldfien, and D. A. Carson.** 1987. High frequency expression of a conserved kappa variable region gene in chronic lymphocytic leukemia. *Proc. Natl. Acad. Sci. USA* **84:**2916–2920.

60. **Kofler, R., F. J. Dixon, and A. N. Theofilopoulos.** 1987. The genetic origin of autoantibodies. *Immunol. Today* **8:**374–380.

61. **Kunkel, H. G., V. Agnello, F. G. Joslin, R. J. Winchester, and J. D. Capra.** 1973. Cross idiotypic specificity among monoclonal IgM proteins with anti-γ-globulin activity. *J. Exp. Med.* **137:**331–342.

62. **Kunkel, H. G., M. Mannik, and R. C. Williams.** 1963. Individual antigenic specificity of isolated antibodies. *Science* **140:**1218–1220.

63. **Kunkel, H. G., R. J. Winchester, F. G. Joslin, and J. D. Capra.** 1974. Similarities in the light chains of anti-γ-globulins showing cross-idiotypic specificities. *J. Exp. Med.* **139**:128–136.

64. **Lee, K. H., F. Matsuda, T. Kinashi, M. Kodaira, and T. Honjo.** 1987. A novel family of variable region genes of the human immunoglobulin heavy chain. *J. Mol. Biol.* **195**:761–768.

65. **Lefvert, A.-K.** 1981. Anti-idiotypic antibodies against the receptor antibodies in myasthenia gravis. *Scand. J. Immunol.* **13**:493–497.

66. **Lefvert, A.-K., R. W. James, C. Alliod, and B. W. Fulpius.** 1982. A monoclonal anti-idiotypic antibody against anti-receptor antibodies from myasthenic sera. *Eur. J. Immunol.* **12**:790–792.

67. **Mageed, R. A., M. R. Walker, and R. Jefferis.** 1986. Restricted light chain subgroup expression on human rheumatoid factor paraproteins determined by monoclonal antibodies. *Immunology* **59**:473–478.

68. **Male, D., G. Pryce, R. Quartey-Papafio, and I. Roitt.** 1983. The occurrence of defined idiotypes on autoantibodies to mouse thyroglobulin. *Eur. J. Immunol.* **13**:942–947.

69. **Marion, T. N., A. R. Lawton III, J. F. Kearney, and D. E. Briles.** 1982. Anti-DNA autoantibodies in (NZB × NZW) F_1 mice are clonally heterogeneous, but the majority share a common idiotype. *J. Immunol.* **128**:668–674.

70. **Matsuyama, T., J. Fukumori, and H. Tanaka.** 1983. Evidence of unique idiotypic determinants and similar idiotypic determinants on human antithyroglobulin antibodies. *Clin. Exp. Immunol.* **51**:381–386.

71. **McLaughlin, C. L., and A. Solomon.** 1972. Bence-Jones proteins and light chains of immunoglobulins. VII. Localization of antigenic sites responsible for immunochemical heterogeneity of kappa chains. *J. Biol. Chem.* **247**:5017–5025.

72. **Mendlovic, S., S. Brocke, Y. Shoenfeld, M. Ben-Bassat, A. Meshorer, R. Bakimer, and E. Mozes.** 1988. Induction of a systemic lupus erythematosus-like disease in mice by a common human anti-DNA idiotype. *Proc. Natl. Acad. Sci. USA* **85**:2260–2264.

73. **Mendlovic, S., H. Fricke, Y. Shoenfeld, and E. Mozes.** 1989. The role of anti-idiotypic antibodies in the induction of experimental systemic lupus erythematosus in mice. *Eur. J. Immunol.* **19**:729–734.

74. **Morgan, A. C., Jr., R. D. Rossen, and J. J. Twomey.** 1979. Naturally occurring circulating immune complexes: normal human serum contains idiotype-anti-idiotype complexes dissociable by certain IgG antiglobulins. *J. Immunol.* **122**:1672–1680.

75. **Naparstek, Y., J. Andre-Schwartz, T. Manser, L. J. Wysocki, L. Breitman, D. B. Stollar, M. Gefter, and R. S. Schwartz.** 1986. A single germline V_H gene segment of normal A/J mice encodes autoantibodies characteristic of systemic lupus erythematosus. *J. Exp. Med.* **164**:614–626.

76. **Neilson, E. G., E. McCafferty, S. M. Phillips, M. D. Clayman, and C. J. Kelly.** 1984. Anti-idiotypic immunity in interstitial nephritis. II. Rats developing anti-tubular basement membrane disease fail to make an anti-idiotypic regulatory response: the modulatory role of an RT 7.1[+], OX8[−] suppressor T cell mechanism. *J. Exp. Med.* **159**:1009–1026.

77. **Neilson, E. G., and M. S. Phillips.** 1982. Suppression of interstitial nephritis by auto-anti-idiotypic immunity. *J. Exp. Med.* **155**:179–189.

78. **Newkirk, M., and J. D. Capra.** 1987. Cross-reactive idiotypic specificity among human rheumatoid factors. *Monogr. Allergy* **22**:1–12.

79. **Newkirk, M., R. A. Mageed, R. Jefferis, P. P. Chen, and D. Capra.** 1987. Complete amino acid sequence of variable regions of two human IgM rheumatoid factors, BOR and KAS of the Wa idiotypic family, reveal restricted use of heavy and light chain variable and joining region gene segments. *J. Exp. Med.* **166**:550–564.

80. **Newkirk, M. M., and J. D. Capra.** 1989. Restricted usage of immunoglobulin variable-region genes in human autoantibodies, p. 203–220. *In* T. Honjo, F. W. Alt, and T. H. Rabbitts (ed.), *The Immunoglobulin Genes.* Academic Press, Inc., New York.

81. **Newkirk, M. M., and J. D. Capra.** 1990. Manuscript in preparation.

82. **Oudin, J.** 1974. L'idiotypie des anticorps. *Ann. Immunol.* (Paris) **125C:**309–337.

83. **Oudin, J., and M. Michel.** 1963. Une nouvelle forme d'allotypie des globulines du serum de lapin apparemment liee a la fonction et a la specificite anticorps. *C.R. Acad. Sci. Ser. D* **257:**805–808.

84. **Pasquali, J. L., A. Urlacher, and D. Storck.** 1983. A highly conserved determinant on human rheumatoid factor idiotypes defined by a mouse monoclonal antibody. *Eur. J. Immunol.* **13:**197–201.

85. **Paul, W. E., and C. Bona.** 1982. Regulatory idiotopes and immune networks: a hypothesis. *Immunol. Today* **3:**230–234.

86. **Pisetsky, D. S., and E. A. Lerner.** 1982. Idiotypic analysis of a monoclonal anti-Sm antibody. *J. Immunol.* **129:**1489–1492.

87. **Pisetsky, D. S., D. M. Salistad, and J. C. Chambers.** 1984. Characterization and idiotypic analysis of an anti-RNP monoclonal antibody. *Clin. Exp. Immunol.* **56:**593–600.

88. **Rauch, J., E. Murphy, J. B. Roths, B. D. Stollar, and R. S. Schwartz.** 1982. A high frequency idiotypic marker of anti-DNA autoantibodies in MRL-lpr/lpr mice. *J. Immunol.* **129:**236–241.

89. **Reininger, L., P. Ollier, P. Poncet, A. Kaushik, and J.-C. Jaton.** 1987. Novel V genes encode virtually identical variable region genes of six murine monoclonal anti-bromelain-treated red blood cell autoantibodies. *J. Immunol.* **138:**316–323.

90. **Rodkey, L. S.** 1974. Studies of idiotypic antibodies. Production and characterization of autoantiidiotypic antisera. *J. Exp. Med.* **139:**712–720.

91. **Rose, L. M., and P. H. Lambert.** 1980. The natural occurrence of circulating idiotype-anti-idiotype complexes during a secondary immune response to phosphorylcholine. *Clin. Immunol. Immunopathol.* **15:**481–492.

92. **Sanz, I., H. Dang, M. Takei, N. Talal, and D. J. Capra.** 1989. V_H sequence of a human anti-Sm autoantibody. *J. Immunol.* **142:**883–889.

93. **Schlomchik, M. J., A. Marshak-Rothstein, C. B. Wolfowicz, T. L. Rothstein, and M. Weigert.** 1987. The role of clonal selection and somatic mutation in autoimmunity. *Nature* (London) **328:**805–811.

94. **Schlomchik, M. J., D. Nemazee, V. L. Sato, J. Van Snick, D. A. Carson, and M. Weigert.** 1986. Variable region sequences of murine IgM anti-IgG monoclonal autoantibodies (rheumatoid factors). *J. Exp. Med.* **164:**407–427.

95. **Schlomchik, M. J., D. Nemazee, J. Van Snick, and M. Weigert.** 1987. Variable region sequences of murine IgM anti-IgG monoclonal auto-antibodies (rheumatoid factors). *J. Exp. Med.* **165:**970–987.

96. **Schreiber, A. B., P. O. Couraud, C. Andre, B. Vray, and A. D. Strosberg.** 1980. Anti-alprenolol anti-idiotypic antibodies bind to anti-adrenergic receptors and modulate catecholamine-sensitive adenylate cyclase. *Proc. Natl. Acad. Sci. USA* **77:**7385–7389.

97. **Schwartz, M., D. Novick, D. Givol, and S. Fuchs.** 1978. Induction of anti-idiotypic antibodies by immunization with syngeneic spleen cells educated with acetylcholine receptors. *Nature* (London) **273:**543–545.

98. **Sege, K., and P. A. Peterson.** 1978. Use of anti-idiotypic antibodies as cell-surface receptor probes. *Proc. Natl. Acad. Sci. USA* **75:**2443–2447.

99. **Seligmann, M., and J. C. Brouet.** 1973. Antibody activity of human myeloma globulins. *Semin. Hematol.* **10:**163–177.

100. **Shechter, Y., R. Maron, D. Elias, and I. R. Cohen.** 1982. Autoantibodies to insulin

receptor spontaneously develop as anti-idiotypes in mice immunized with insulin. *Science* **216**:542–545.

101. **Shen, A., C. Humphries, P. Tuker, and F. Blattner.** 1987. Human heavy chain variable region gene family nonrandomly rearranged in familial chronic lymphocytic leukemia. *Proc. Natl. Acad. Sci. USA* **84**:8563–8567.

102. **Shoenfeld, Y., D. A. Isenberg, J. Rauch, M. P. Madaio, B. D. Stollar, and R. S. Schwartz.** 1983. Idiotypic cross-reactions of monoclonal human lupus autoantibodies. *J. Exp. Med.* **158**:718–730.

103. **Silverman, G. J., D. A. Carson, K. Patrick, J. H. Vaughan, and S. Fong.** 1987. Expression of germline human kappa chain associated cross-reactive idiotype after in vitro and in vivo infection with Epstein-Barr virus. *Clin. Immunol. Immunopathol.* **43**:403–411.

104. **Silverman, G. J., R. D. Goldfien, P. P. Chen, R. A. Mageed, R. Jefferis, F. Goni, B. Frangione, S. Fong, and D. A. Carson.** 1988. Idiotypic and subgroup analysis of human monoclonal rheumatoid factors. Implication for structural and genetic bases of autoantibodies in humans. *J. Clin. Invest.* **82**:469–475.

104a.**Sollazzo, M., D. Castiglia, R. Billetta, A. Tramontano, and M. Zanetti.** 1990. Structural definition by antibody engineering of an idiotypic determinant. *Protein Eng.,* submitted for publication.

105. **Sollazzo, M., C. A. Hasemann, K. D. Meek, D. Glotz, J. D. Capra, and M. Zanetti.** 1989. Molecular characterization of the V_H region of murine autoantibodies from neonatal and adult BALB/c mice. *Eur. J. Immunol.* **19**:453–457.

106. **Solomon, A., and C. L. McLaughlin.** 1971. Bence-Jones proteins and light chains of immunoglobulins. IV. Immunochemical differentiation among proteins within each of the three established kappa chain classes. *J. Immunol.* **106**:120–127.

107. **Solomon, G., J. Schiffenbauer, H. D. Keiser, and B. Diamond.** 1983. Use of monoclonal antibodies to identify shared idiotypes on human antibodies to native DNA from patients with systemic lupus erythematosus. *Proc. Natl. Acad. Sci. USA* **80**:850–854.

108. **Tachovsky, T. G., M. Sandberg-Wollheim, and L. G. Baird.** 1982. Rabbit anti-human CSF IgG. I. Characterization of anti-idiotype antibodies produced against MS CSF and detection of cross-reactive idiotypes in several MS CSF. *J. Immunol.* **129**:764–770.

109. **Teitelbaum, D., J. Rauch, B. D. Stollar, and R. S. Schwartz.** 1984. In vivo effects of antibodies against a high frequency idiotype of anti-DNA antibodies in MRL mice. *J. Immunol.* **132**:1282–1285.

110. **Thomas, J. W., V. J. Virta, and L. J. Nell.** 1986. Idiotypic determinants on human anti-insulin antibodies are cyclically expressed. *J. Immunol.* **137**:1610–1615.

111. **Trepicchio, W., Jr., A. Maruya, and K. J. Barret.** 1987. The heavy chain genes of a nonautoimmune strain of mouse are conserved in strains of mice polymorphic for this gene locus. *J. Immunol.* **139**:3139–3145.

112. **Tron, F., C. Le Guern, P.-A. Cazenave, and J.-F. Bach.** 1982. Intrastrain recurrent idiotypes among anti-DNA antibodies of (NZB × NZW) F_1 hybrid mice. *Eur. J. Immunol.* **12**:761–766.

113. **Wager, O., and A. M. Teppo.** 1978. Binding affinity of human auto-antibodies: studies of cryoglobulin IgM rheumative factors and IgG autoantibodies to albumin. *Scand. J. Immunol.* **7**:503–509.

114. **Wasserman, N. H., A. S. Penn, P. I. Freimuth, N. Treptow, S. Wentzel, W. L. Cleveland, and B. F. Erlanger.** 1983. Anti-idiotypic route to anti-acetylcholine receptor antibodies and experimental myasthenia gravis. *Proc. Natl. Acad. Sci. USA* **79**:4810–4814.

115. **Wedgewood, J., N. Ilowite, A. Lewison-Nisen, and V. Bonagura.** 1987. Seronegative JRA patients express the major rheumatoid factor cross-reactive idiotype. *Pediatr. Res.* **21**:320A. (Abstract no. 878.)

116. **Wekerle, H.** 1978. Immunological T-cell memory in the in vitro-induced experimental orchitis. *J. Exp. Med.* **147**:233–250.
117. **Williams, R. C., H. G. Kunkel, and J. D. Capra.** 1968. Antigenic specificities related to the cold agglutinin activity of gamma M globulins. *Science* **161**:379–380.
118. **Zanetti, M., and P. E. Bigazzi.** 1981. Anti-idiotypic immunity and autoimmunity. I. In vitro and in vivo effects of anti-idiotypic antibodies to spontaneously occurring autoantibodies to rat thyroglobulin. *Eur. J. Immunol.* **11**:187–195.
119. **Zanetti, M.** 1983. Anti-idiotypic antibodies and autoantibodies. *Ann. N.Y. Acad. Sci.* **418**:363–387.
120. **Zanetti, M.** 1985. The idiotypic network in autoimmune processes. *Immunol. Today* **6**:299–302.
121. **Zanetti, M.** 1986. Idiotype network and its relevance to autoimmune diseases. Functional considerations. *Concepts Immunopathol.* **3**:253–284.
122. **Zanetti, M.** 1988. Self immunity and the autoimmune network: a molecular perspective to ontogeny and regulation of the immune system. *Ann. Immunol. Inst. Pasteur* **139**:619–631.
123. **Zanetti, M., R. W. Barton, and P. E. Bigazzi.** 1983. Anti-idiotypic immunity and autoimmunity. II. Idiotypic determinants of autoantibodies and lymphocytes in spontaneous and experimentally induced autoimmune thyroiditis. *Cell. Immunol.* **75**:292–299.
124. **Zanetti, M., R. Glotz, and J. Rogers.** 1986. Perturbation of the autoimmune network. II. Immunization with isologous idiotype induces autoantiidiotypic antibodies and suppresses the immune response elicited by antigen. *J. Immunol.* **137**:3140–3146.
125. **Zanetti, M., F.-T. Liu, J. Rogers, and D. Katz.** 1985. Heavy and light chains of a mouse monoclonal autoantibody express the same idiotype. *J. Immunol.* **135**:1245–1251.
126. **Zanetti, M., F. Mampaso, and C. B. Wilson.** 1983. Anti-idiotype as a probe in the analysis of autoimmune tubulointerstitial nephritis in the Brown Norway rat. *J. Immunol.* **131**:1268–1273.
127. **Zanetti, M., and J. Rogers.** 1987. Independent expression of a regulatory idiotype on heavy and light chains. A further immunochemical analysis with anti-heavy and anti-light chain antibodies. *J. Immunol.* **139**:1965–1970.
128. **Zanetti, M., J. Rogers, and D. H. Katz.** 1984. Induction of autoantibodies to thyroglobulin by anti-idiotypic antibodies. *J. Immunol.* **133**:240–243.
129. **Zanetti, M., and C. B. Wilson.** 1983. Participation of auto-anti-idiotypes in immune complex glomerulonephritis in rabbits. *J. Immunol.* **131**:2781–2783.
130. **Ziff, M.** 1985. Autoimmune aspects of rheumatoid arthritis, p. 59–79. *In* N. R. Rose and I. R. Mackay (ed.), *The Autoimmune Diseases.* Academic Press, Inc., New York.

Chapter 9

Modulation of Antitumor Immunity by Anti-Idiotypic Antibodies

Martine Wettendorff, Hilary Koprowski, and Dorothee Herlyn

The existence of tumor-associated antigens (TAAs) was first demonstrated by Foley et al. (18). The development in the last decade of monoclonal antibodies has greatly contributed to the characterization of TAAs as well as to the immune responses they induce. Three classes of TAAs have been defined (78). Class I TAAs are expressed on an individual tumor and not on other tumors of the same histological type or on normal cells. They are rare. Class II TAAs are expressed on different tumors of the same histological type derived from different individuals but are not expressed on normal cells. The most common TAAs belong to the class III. They are found on malignant cells and normal cells but are usually expressed at lower density on normal cells. Since antigen density on cell surfaces has been shown to correlate with cell destruction by antibodies (35), active and passive immunotherapy of cancer patients with class III TAAs (with high expression on tumor cells, but low expression on normal cells) has not produced any adverse side effects that could have been related to the expression of the TAAs by normal cells (41, 96).

In approaches to active immunotherapy of cancer, autologous irradiated tumor cells, purified TAAs, and anti-idiotypic antibodies (Ab_2) bearing internal images of TAAs have been used (reviewed in reference 40). Each of these compounds has advantages and disadvantages. The specificity of the reagents is highest for Ab_2, since they may mimic a single determinant of TAA, whereas TAAs include multiple determinants of an antigen and tumor cells express numerous potentially immunogenic TAAs. Since the immunogenicity of antigenic compounds is directly correlated with the

---

Martine Wettendorff, Hilary Koprowski, and Dorothee Herlyn   •   The Wistar Institute, 36th Street at Spruce, Philadelphia, Pennsylvania 19104.

number of antigenic determinants, tumor cells display the highest immunogenicity, followed by TAA and Ab_2. However, in immunologically tolerant hosts, including cancer patients, who are usually tolerant to TAAs (30, 113), Ab_2 may induce immunity whereas TAA may not, analogous to the demonstration of breakage of immunological tolerance to polysaccharide antigen by Ab_2 immunizations of neonatal mice (104). Ab_2 as vaccine candidates are safe, whereas tumor-derived compounds bear the intrinsic danger of viral contamination. Finally, Ab_2, especially monoclonal Ab_2, are easy to produce in large quantities, whereas autologous tumor cells may not easily grow in continuous culture, and TAA is extremely difficult to purify in large quantities unless it is available in recombinant form. However, to our knowledge, recombinant TAA has not been used in clinical trials.

IDIOTYPIC CASCADES INITIATED BY TUMOR ANTIGEN AND ANTITUMOR ANTIBODY (Ab_1)

TAAs expressed by growing tumors may induce idiotypic networks similar to idiotypic cascades demonstrated in humans after exposure to non-tumor-derived antigens (1, 2, 13, 23–25, 59, 64, 94, 100). Figure 1 shows the different idiotypic pathways that can be initiated by stimulation of the immune system with TAA. TAA may induce idiotype/antigen receptor-positive T_H cells which help B cells produce idiotype/antigen receptor-positive antibody (Ab_1). Id^+ T_H cells and/or Ab_1 may induce Ab_2 which either suppress (via induction of Id^+ T_S cells) or activate Id^+ ($Ab_1{}^+$) B cells. It is also conceivable that TAA directly induces the formation of idiotype (antigen receptor)-positive T_S cells, which suppress the formation of TAA-specific Ab_1 and/or induce the production of T_S cell-specific Ab_2. These Ab_2 may down regulate Ab_1 production.

Circulating antibodies (Ab_1) to TAAs have been found in cancer patients, although this represents a rare event (reviewed in reference 33). Even in the presence of antitumor responses, the tumor usually is able to escape the immunological surveillance system. The specific escape mechanisms

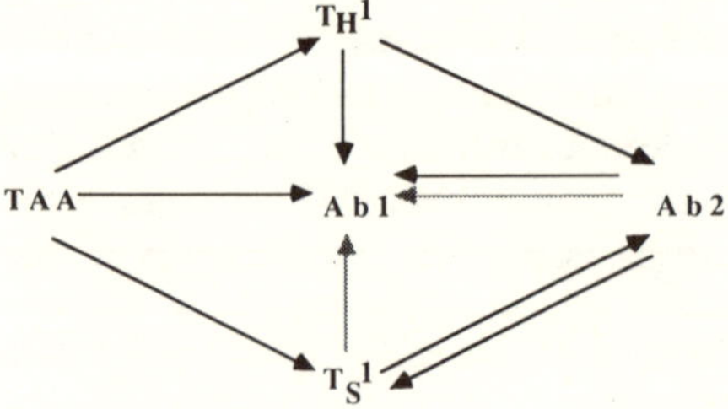

Figure 1. Idiotypic and anti-idiotypic responses initiated by TAA. →, Induction pathway; →, suppression pathway; $T_H{}^1$, $T_S{}^1$, idiotype-positive helper and suppressor T cells, respectively.

remain unclear, although several possibilities have been discussed. Down regulation of some TAAs in the presence of antibody has been described (10, 88). This modulation includes mechanisms such as shedding, endocytosis, or redistribution of the TAAs on the cell membrane. Tumor heterogeneity may also influence the immune responses. Immune responses against heterogeneous tumors have been shown to select for the less immunogenic cells (68).

Since TAAs are immunologically cross-reactive with normal tissue antigens, this could explain why cancer patients are ordinarily tolerant of TAAs (30, 113). The absence of antibody induction to TAAs may be the result of specific suppression of Id^+ B cells by regulatory molecules such as Ab_2 or T_S factors (Fig. 1). Id^+ T_S cells as well as Id^+ T_H cells have been demonstrated in response to TAA stimulation (31, 72, 73). In the mouse sarcoma system, Id^+ T_S cells were generated spontaneously during the antitumor response (74). The evidence that these cells carry idiotopes was given by the demonstration that the cells interact with Ab_2. In the mouse sarcoma system, T_S cells could have substituted for the absent Ab_1 and served as the stimulus for the generation of Ab_2 in an idiotypic cascade, TAA $\rightarrow$ Id^+ T_S cell $\rightarrow$ Ab_2.

We have demonstrated idiotypic cascades in cancer patients undergoing passive immunotherapy with monoclonal antitumor antibody (Ab_1 [54, 113]). Gastrointestinal cancer patients treated with Ab_1 CO17-1A developed Ab_2 to the Ab_1 (54, 113). The Ab_2 functionally mimicked in vitro the antigen defined by monoclonal antibody (MAb) CO17-1A (40, 54). We had originally postulated that Ab_3 with Ab_1-like binding specificities to tumor cells are induced by Ab_2 (54). We further postulated that Ab_3 responses may underlie the clinical responses observed in some of the MAb-treated patients (96). In the initial studies, Ab_3 could not be demonstrated in patient sera, presumably because the Ab_3 were bound to circulating Ab_2 or to patient tumor cells. Recently, however, Ab_3 were demonstrated in culture supernatants of peripheral-blood mononuclear cells (PBMC). Cells from six Ab_1-treated patients were stimulated with heterologous Ab_2 that functionally mimicked the TAA defined by Ab_1 and were immunologically cross-reactive with the patient Ab_2 (113) (Fig. 2). Ab_3 bound to autologous Ab_2, shared idiotopes with Ab_1, and were Ab_1-like in their binding specificities to tumor cells, TAA, and Ab_2. Antibodies with identical binding specificities to those of Ab_3 were also elicited by stimulating PBMC with isolated TAA. However, the antibodies were not produced by stimulating posttreatment PBMC with control proteins or by stimulating pretreatment cells with either TAA or Ab_2; this suggests that B cells of cancer patients are tolerant to the CO17-1A antigen expressed by the growing tumors (113) (Fig. 2). Our results demonstrate the presence of idiotypic cascades in cancer patients

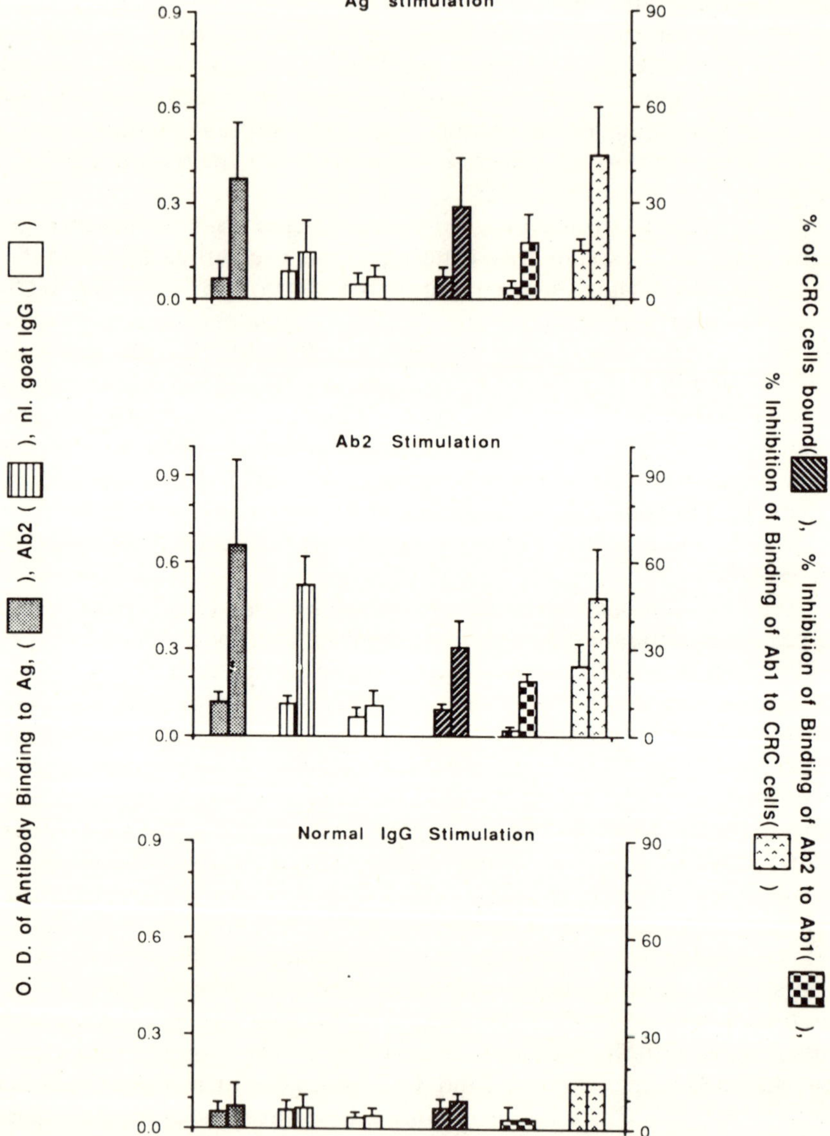

Figure 2. In vitro stimulation of PBMC from pancreatic cancer patients treated with monoclonal antitumor antibody; demonstration of idiotypic cascades. PBMC were obtained from patients before (first column of each set of two columns) or after (second column of each set of two columns) therapy with MAb CO17-1A. PBMC were stimulated with optimal concentrations (between 1 and 20 ng/ml) of either antigen, Ab_2, or normal goat immunoglobulin G or left untreated. Supernatants derived from each set of PBMC were tested for binding to antigen,

treated with Ab_1. Although it is not possible, because of the small number of patients in the study, to relate immunological responses of the patients to clinical antitumor responses, preliminary data obtained with PBMC from one colon carcinoma patient whose tumor completely regressed after Ab_1 therapy showed significant antigen-specific Ab_3 responses.

Antibodies binding to antigens were also found in sera of patients with ovarian cancer that had been treated with repeated injections of radiolabeled Ab_1 (11). It has been proposed (11) that Ab_2 found in the sera of the treated patients may have carried an internal image of the TAA and induced antibodies with Ab_1-like activity. However, it is unclear from the results of that study whether these antibodies are indeed Ab_3 or, rather, may represent Ab_1, since pretherapy sera were not included in the study.

The induction of idiotypic cascades by Ab_1 immunizations in animal models has been described for a virus-induced TAA (81). In that system, injection of Ab_1 was followed by the production of both Ab_2 and Ab_1-like Ab_3. Surprisingly, Ab_3 was present in sera of mice in much higher concentrations than Ab_1 induced by virus injection. The Ab_3 may have played a role in tumor regression observed in Ab_1-treated mice challenged with tumor cells.

MANIPULATION OF IDIOTYPIC RESPONSES TO TAA BY ADMINISTRATION OF Ab_2

In the development of Ab_2 as an immunotherapeutic agent against cancer, careful selection of antitumor antibody (Ab_1) and Ab_2 and optimization of the Ab_2 administration schedule are most critical.

Selection of Ab_1

Selection of Ab_1 is based on several criteria: (i) Ab_1 specificity; (ii) immunodominance of the TAA defined by Ab_1; (iii) cytotoxic characteristics of Ab_1; and (iv) availability of the TAA.

Class III TAAs are usually expressed at a much higher density on tumor cells than on normal cells. Therefore, most Ab_1 bind preferentially to tumor cells, with low binding to normal cells. Such Ab_1 may be selected for production of Ab_2 despite the expression of the TAA by normal cells (see the introduction to this chapter).

In cancer patients, immunodominance of the epitope defined by the

---

Ab_2, normal goat immunoglobulin G, or colon carcinoma cells SW 1116, for idiotope sharing with the Ab_1 CO17-1A, and for epitope specificity. Values represent means and standard deviations of PBMC supernatants from five patients.

Ab_1 facilitates the induction of immunity by an Ab_2 mimicking this epitope. However, even in the absence of B-cell reactivity to a TAA expressed by gastrointestinal carcinomas, a state of active immunity against this TAA could be induced in the patients by immunization with Ab_2 (41).

Ab_2 against tumoricidal Ab_1 may induce the production of Ab_3 that not only mimic the binding specificities of Ab_1 but also exhibit tumoricidal activity. However, cytotoxic activity of the Ab_3 depends on parameters such as (i) the isotype of the elicited Ab_3, (ii) the number of Ab_3 molecules bound to Fc receptors on effector cells, and/or (iii) the binding affinity of the Ab_3 to these receptors and to antigen (32, 34).

The characterization of Ab_2 that is thought to carry the internal image of the antigen would be facilitated if the antigen itself were available (36, 37).

Selection of Ab_2

Immunotherapy with Ab_2 is based on two different concepts, depending on the type of Ab_2 used (Fig. 3). Ab_2 directed against regulatory idiotopes of Ab_1 ($Ab_2\alpha$ [6]) participate in the regulatory network of the antitumor response. These Ab_2 bind to idiotopes outside the antigen-combining region of Ab_1 and thus do not inhibit binding of the Ab_1 to the antigen. The expression of regulatory idiotopes may be closely linked to the antigen-binding specificity of the Ab_1; however, regulatory idiotope expression may also be linked to unknown (unwanted) specificities (6), and their expression usually is genetically restricted. Therefore, Ab_2 to regulatory idiotopes may induce antigen-binding Ab_3 only in certain species or strains of animals (19, 91). Forstrom et al. (19) showed that Ab_2 immunization elicited delayed-type hypersensitivity (DTH) reaction to a murine sarcoma in an Igh-restricted fashion; i.e., only mice expressing the same immunoglobulin allotype as the Ab_2 were able to respond to Ab_2 immunization.

In contrast to Ab_2 against regulatory idiotopes, internal-image Ab_2 ($Ab_2\beta$ [Fig. 3]) can act as surrogate antigen (77). Jerne defined an internal-image Ab_2 ($Ab_2\beta$) as an Ab_2 carrying an idiotope that resembles the epitope

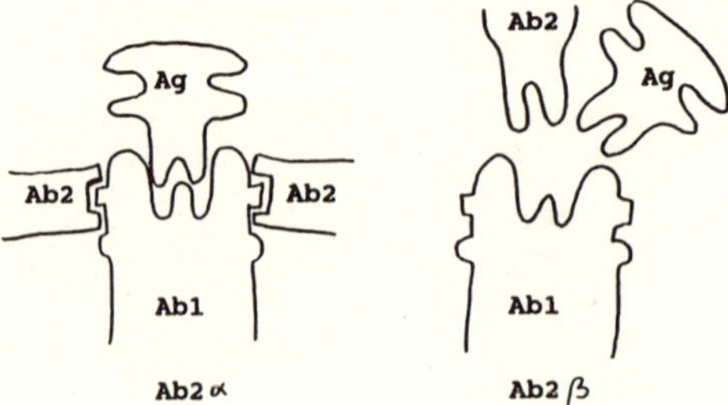

Figure 3. Binding specificities of $Ab_2\alpha$ and $Ab_2\beta$.

of the antigen and therefore binds to the combining site of the Ab_1 (45). These Ab_2 are also called homobodies. Figure 3 compares the binding specificities of $Ab_2\alpha$ and $Ab_2\beta$. $Ab_2\alpha$ binds to regulatory idiotypes, but only the $Ab_2\beta$ can act as surrogate of the antigen and is therefore directly comparable to the immunogen itself. $Ab_2\beta$ mimic functional (98, 109) and structural (8, 79, 111) characteristics of the antigen. Criteria to define $Ab_2\beta$ are well established (15). They are as follows. (i) The most useful criterion to define $Ab_2\beta$ is its capacity to functionally mimic antigen, as originally demonstrated in the insulin (98) and tobacco mosaic virus (109) systems. Immune responses to $Ab_2\beta$ should be no more restricted genetically than immune responses induced by an antigen. Therefore, induction of antigen-specific immunity by Ab_2 immunizations across species barriers has been most widely used as a criterion to identify $Ab_2\beta$. (ii) The structural identity of a segment of the variable region of the heavy and light chain of an Ab_2 and the sequence of a given epitope is the most rigorous criterion to define an internal-image Ab_2 (8, 79, 111). However, this criterion is applicable only to protein structures, and many protein sequences are not available. (iii) Immunochemical criteria are also useful. $Ab_2\beta$ should bind to the homologous Ab_1 as well as to heterologous antibodies directed against the same epitope. The binding of $Ab_2\beta$ to the Ab_1 should be inhibited by purified antigen. However, in the latter case, steric hindrance of antigen-combining regions of Ab_1 by antigen may obscure the interpretation of the results.

Figure 4 shows the different pathways induced by Ab_2 immunizations. Ab_2 can select Id^+ B cells which may produce Id^+ antibody (Ab_3). Alternatively, Ab_2 may induce Id^+ T_H cells (T_H3), which could help B cells in the production of Ab_3; or Ab_2 may induce T_S cells (T_S3) or cytotoxic T lymphocytes (CTL). Although in each case Ab_3 or T cells share idiotopes with Ab_1, only a fraction of Id^+ Ab_3 and Id^+ T cells may bind antigen when polyclonal Ab_2 are used for immunization, whereas in immunizations with monoclonal Ab_2 of the internal-image type (see above), each of the Ab_3 molecules induced by Ab_2 has the potential of binding the antigen.

Ab_2 have been shown to induce antigen-specific humoral and cellular

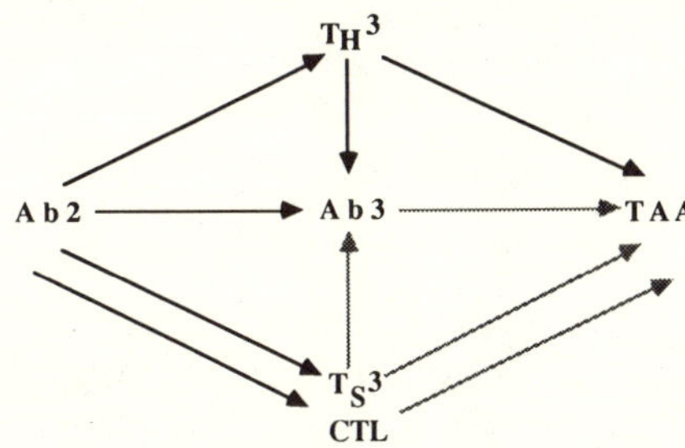

Figure 4. Induction of antitumor immunity by Ab_2. →, Induction pathway; →, binding reaction; T_H^3, T_S^3, idiotype-positive helper and suppressor T cells, respectively.

immune responses in experimental models including various tumor, virus, and bacterial systems (reviewed in references 39, 53, 75, 91, and 101). The induction of B-cell immunity by Ab_2 differs fundamentally from the induction of T-cell immunity. B cells preferentially recognize conformational epitopes, whereas T_H cells recognize small antigenic fragments of processed antigen in association with class II major histocompatibility complex (MHC) molecules (63, 95). CTL most probably recognize endogenously produced antigen in association with class I MHC molecules (108). Shared idiotopes between B and T cells specific for the same antigen provide the basis for induction of both humoral and cellular antigen-specific immune responses by Ab_2 directed to B-cell idiotopes (22, 76, 83–85, 99, 103). Several groups showed that Ab_2 to B-cell (Ab_1) idiotopes can induce T_H cells and CTL in chemically and virally induced tumors (19, 57, 73, 74, 83, 85, 103).

Stimulation in vitro (and most probably also in vivo) of T cells by Ab_2 (directed against immunoglobulin idiotopes) usually requires processing and presentation of the Ab_2 by B cells and/or macrophages in the context of MHC molecules (86, 87, 102). In contrast to Ab_2 against immunoglobulin idiotopes, anti-idiotypes to T-cell idiotopes (clonotypic antibodies) have been shown to stimulate T_H cells directly without the context of MHC molecules (44, 49), suggesting that the antibodies may bind antigen and/or MHC recognition sites on the T cell. Most interestingly, clonotypic antibodies to T_H cell clones were shown to induce antigen-specific T cells and protective immunity against a subsequent infection with virus or bacteria across H-2 barriers (17, 48). In one of those studies (17), CTL, induced by clonotypic antibodies directed to T_H cell clones, lysed virus-infected target cells irrespective of their H-2 haplotypes, whereas antigen-induced CTL lysed only syngeneic virus-infected target cells. It is unclear why induction of antigen-specific cellular immunity by antigen or Ab_2 to immunoglobulin idiotopes differs from the immunity induced by Ab_2 to T-cell idiotopes in both accessory cell dependency and MHC restriction. Anti-idiotypes against T-cell idiotopes may bind with higher affinity to antigen-specific T cells than to antigen, thereby obliterating the need for concomitant MHC recognition by Ab_2-stimulated T cells (16). From these results, obtained by using viral and bacterial systems (17, 48), Ab_2 against T-cell idiotopes may represent prime candidates for inducers of cellular antitumor immunity, although no such antibodies have been described thus far.

Ab_2 Administration Schedule

A systematic study of the conditions that best induce maximum antigen-specific responses to Ab_2 immunizations has not been performed yet. In a hapten system described by Kelsoe et al. (50) the dose of Ab_2 seems to be very important. The immunogenicity of Ab_2 may be increased by the use of adjuvants (41, 67), by coupling of the Ab_2 to lipopolysaccharide (20), or by presentation of the Ab_2 by antigen-presenting cells (21). The route of immunization has also been shown to influence the immune response. In studies with hapten proteins (29), intraperitoneal immunization with Ab_2 preferentially activated T_H cells, whereas intravenous immunization induced T_S cells.

Another important factor that affects the outcome of Ab_2 immunizations is the form of Ab_2 administered. In xenogeneic immunizations with Ab_2, intact Ab_2 molecules will elicit strong anti-iso/allotypic immune responses which do not interact with antigen, as demonstrated in cancer patients treated with goat Ab_2 (41). Although anti-iso/allotypic T_H cells may help B cells respond to the internal-image portion of the Ab_2 (92, 113), these T_H cells are of no direct therapeutic (tumoricidal) value. Replacement of the Fc by a more useful carrier, such as tetanus toxoid, may be advantageous in approaches to Ab_2 immunizations of humans. Uncoupled $F(ab')_2$ fragments of Ab_1 have been shown to induce Ab_2 responses (107), and therefore, fragments of Ab_2 may provide an alternative approach to the use of intact Ab_2. Furthermore, it has been observed that intact Ab_2 in some instances induced antigen-specific T_S cells, whereas fragments of the same Ab_2 did not (29). Human monoclonal Ab_2 directed against Ab_1 to human tumor-associated antigen have been proposed as vaccine candidates for cancer patients (7, 105), although such antibodies bear the intrinsic hazard of viral contamination of human B-cell-derived antibodies.

The above-described factors that may influence the immune responses induced by Ab_2 immunizations should be carefully studied in animal models.

Induction of Immunity to Murine Tumors by Ab_2 Immunizations

Table 1 summarizes Ab_2 immunizations in different murine tumor systems. One of the first Ab_2 was described in the BALB/c sarcoma system by Forstrom et al. (19). Hyperimmunization of mice with sarcoma cells was used to produce a MAb (antibody 4.72) that induces allotype-restricted DTH specific for sarcoma cells. This antibody did not bind to antigen and can therefore be considered an Ab_2 even in the absence of a defined Ab_1. Because of the allotype restriction of DTH induced by this antibody, it is probably not an $Ab_2\beta$.

More recent studies by Nelson et al. (73) showed that the idiotope defined by antibody 4.72 (Ab_2) was expressed on B cells of BALB/c mice responding to the sarcoma cells and also on products of T cells that suppressed the tumor-specific DTH response. From these data, a mechanism was proposed that could underlie tumor growth inhibition in mice immu-

Table 1. Induction of Immunity to Murine and Rat Tumors by Ab_2 Immunizations[a]

Origin of tumor	Ab_1			Ab_2		
	Induced in:	Clonality	Binding specificity	Induced in:	Clonality	Binding specificity
Sarcoma (mouse)		NA		BALB/c mouse	Mono-clonal	NT (because of unavailability of Ab_1)
Transit cell bladder carcinoma (mouse)	Rat	Mono-clonal	Glycoprotein p175	BALB/c mouse	Mono-clonal	Combining site-associated idiotype on Ab_1
Sarcoma (rat)	Rat	Mono-clonal	Sarcoma cells	Rat	Mono-clonal	Combining site-associated idiotype on homologous Ab_1
SV40-transformed cells (mouse)	BALB/c mouse	Mono-clonal	Large T antigen from SV40	Rabbit	Poly-clonal	Combining site-associated private idiotype on homologous Ab_1 only
Lymphoma (mouse)	BALB/c mouse	Mono-clonal	Glycoprotein gp52 cross-reactive with MMTV	A/SE mouse	Mono-clonal	Combining site-associated idiotype on Ab_1

(*Continued on next page*)

nized with Ab_2. This model is based on the assumptions that B and T cells interact in idiotypic networks. Ab_2 4.72 produced in BALB/c mice during the immune response to sarcoma may inhibit tumor growth through induction of DTH and/or inactivation of T_S-cell factors and thereby define an idiotope expressed on tumor-specific regulatory T cells.

Lee et al. have generated a rat monoclonal antibody that recognizes the oncofetal cell surface antigen p175 on mouse bladder carcinoma (57). Mouse monoclonal Ab_2 were produced and shown to be specific for different idiotopes on the Ab_1, all of which are associated with the binding site for the antigen. The Ab_2 induced a DTH specific for bladder carcinoma in syngeneic mice as well as Ab_3 that bound to idiotopes on Ab_2 but lacked binding reactivity to antigen p175. Induction of cell-mediated immunity by Ab_2 was also tested in vitro by the leukocyte adherence inhibition reaction. In this assay, leukocytes from Ab_2-immunized mice reacted with purified

Table 1—*Continued*

Immunity induced by Ab_2				References
Induced in:	Humoral/cell mediated	Specificity	Effect on challenge with tumor	
BALB/c mouse	DTH	Specific for sarcoma	Inhibition of tumor growth	19, 73
BALB/c mouse	DTH	Specific for bladder carcinoma	NT	57, 58
	Ab_3	Binding to Ab_2 only, not to antigen		
Rat	DTH	Specific for sarcoma	Partial reduction of lung colonies of tumor cells	17
	Ab_3 (polyclonal)	Same binding specificity as Ab_1		
Balb/c mouse	Ab_3 (polyclonal)	Binding to Ab_2 only	Partial inhibition of tumor growth of SV40-transfected cells in vivo	51
DBA/2 mouse	Ab_3 (polyclonal) DTH	Binding to MMTV Specific for lymphoma	Partial inhibition of tumor growth, prolongation of survival	83–86, 92
	CTL	Specific for lymphoma		

[a] Abbreviations: NA, not available; NT, not tested; MMTV, mouse mammary tumor virus; SV40, simian virus 40.

bladder carcinoma antigen. Since the effect of Ab_2 immunizations against a challenge with tumor cells was not tested, it is unknown whether the described Ab_2 have protective potential. An important question raised by this study is why no antigen-binding Ab_3 were produced by Ab_2-immunized mice. The authors suggested (58) that the concentration of antigen-binding Ab_3 could be too low to be detected in the serum, or that B cells of mice are incapable of producing antibodies against the epitope defined by the Ab_1. An incapability might be genetically determined or might be the result of active immune suppression.

Dunn et al. (14) prepared syngeneic rat monoclonal Ab_2 that seemed both in vitro and in vivo to mimic an antigen present on the surface of a chemically induced murine sarcoma. The Ab_2 specifically inhibited the binding of labeled Ab_1 to antigen and induced an Ab_3 with a binding specificity similar to the Ab_1 in rats. The potential of the Ab_2 to protect rats against tumor challenge was also investigated. Immunizations of rats with various doses of Ab_2 either in phosphate-buffered saline or in complete Freund adjuvant were followed by intravenous injections of live tumor cells. Immunizations of rats with high doses of Ab_2 in phosphate-buffered saline induced a significant reduction of lung tumor colonies only.

Kennedy et al. generated polyclonal rabbit Ab_2 against the Ab_1 to large T antigen from simian virus 40-transformed BALB/c mouse cells (51). The Ab_2 recognized a combining-site-associated private idiotope expressed exclusively by the homologous Ab_1. Immunizations of BALB/c mice with the Ab_2 induced the production of Ab_3 that did not bind to large T antigen but did inhibit binding of homologous Ab_2 to Ab_1 (and thus shared idiotopes with Ab_1). Although the Ab_3 did not show any antitumor activity, tumor regression was observed when Ab_2-immunized mice were challenged with tumor cells. The investigators proposed that the Id^+- and antigen-nonbinding Ab_3 may have immune regulatory functions which affect tumor growth. Alternatively, Ab_2 may induce cellular antitumor responses. Such responses have not been investigated in that system.

Raychaudhury et al. reported that the immunization of mice with monoclonal Ab_2 induces specific humoral and cellular immune responses to murine lymphoma (83–85). They produced two Ab_2 in BALB/c mice against the paratope of an antitumor MAb that recognizes a lymphoma-associated antigen which expressed determinants cross-reactive with the mouse mammary tumor virus glycoprotein gp52. Mice immunized with Ab_2 produced Ab_3 with binding specificities for gp52, as well as DTH reactions and CTL specific for lymphoma. Although both Ab_2 were able to induce these immune responses, only one Ab_2 prolonged the survival of tumor-challenged mice. The authors suggested that different populations of T cells may have been activated by the different Ab_2 preparations. To investigate this point further, they produced T_H cell clones against the idiotopes of the nonprotective Ab_2 (93). These clones responded to Ab_2 stimulation but not to stimulation with nominal antigen. Since the Ab_2 induced antigen-specific Ab_3 responses in the mice in the absence of antigen-specific T_H cells (83, 85), this suggests that the Ab_3 responses were T_H cell independent. Furthermore, in light of the demonstration of cytolytic functions of T_H cells in the mouse (12), the lack of T_H-cell induction by this Ab_2 may correlate with the failure of the Ab_2 to induce protective effects.

The studies of animal tumor systems have shown that Ab_2 immunizations induce specific and protective humoral and/or cellular immunity. T-cell responses have been more frequently induced in those systems than B-cell responses. Murine tumor models offer ideal tools for the study of the role of the complex interactions between Ab_2 and either T_H cells, T_S cells, or CTL in the regulation of tumor growth. Such studies are extremely difficult to perform in experimental human tumor systems (see Experimental Induction of Immunity to Human Tumors by Ab_2 Immunizations).

Experimental Induction of Immunity to Human Tumors
by Ab_2 Immunizations

Although investigators in several laboratories have been studying experimental induction of immunity to human tumors by Ab_2 immunizations, these studies must be interpreted cautiously, since an animal model in which one could evaluate the protective effects of Ab_2 immunizations against a challenge with human tumor cells is not readily available. Human tumors grow only in immunocompromised animals which cannot be effectively immunized, although the newly developed SCID mice may offer a suitable model (70). Furthermore, the possibility that Ab_2 immunizations induce CTL reactive to human tumors, as shown in the murine lymphoma system (83), cannot be investigated experimentally, since CTL reactivities are generally restricted to autologous or heterologous MHC-matched targets. Induction of CTL by Ab_2 mimicking human tumor-associated antigen must therefore be investigated directly in cancer patients (see Clinical Trials with Ab_2).

Table 2 summarizes experimental induction in animals by Ab_2 immunizations of specific immunity against human tumors. Nepom et al. (76) raised in rabbits polyclonal Ab_2 to a murine Ab_1 specific for the human melanoma-associated antigen p97. Binding of Ab_1 to Ab_2 was significantly inhibited by purified tumor antigen. The Ab_2 also inhibited the binding of Ab_1 to the antigen. These results suggest that a portion of the polyclonal Ab_2 was directed to the antigen-binding site of the Ab_1 or to an idiotope very close to this site. Sera from BALB/c mice immunized with these Ab_2 contained significant amounts of Ab_3 that inhibited the binding of Ab_2 to Ab_1. Some Id^+ Ab_3-containing sera immunoprecipitated radiolabeled p97 from melanoma cell lysates. Ab_2 also induced a DTH response to a challenge with melanoma cells in mice. It is unclear from this study (76) whether the Ab_2 is of the internal-image type, since induction of immunity by the Ab_2 has not been investigated across species barriers (both Ab_1 and Ab_3 were derived from BALB/c mice). Furthermore, the demonstration that Ab_2 reacts with Ab_1 of various species directed against the same epitope on antigen p97 might clarify this point. Lastly, studies of the immunotherapeutic capabilities of the Ab_3 induced in that system, such as demonstration of cytotoxic reactivities, would be of great interest.

We have developed goat Ab_2 against two MAbs (Ab_1 CO17-1A and GA733) to human gastrointestinal tumor-associated antigen (36, 37, 42). The Ab_1 were selected on the basis of their tumor specificity and their capabilities to mediate cytotoxicity with murine macrophages in vitro and to inhibit growth in vivo in nude mice and cancer patients (32, 34, 38, 96).

Table 2. Induction of Immunity to Human Tumors by Ab_2 Immunizations[a]

Origin of tumor	Ab_1			Ab_2			Immunity induced by Ab_2			References
	Induced in:	Clonality	Binding specificity	Induced in:	Clonality	Binding specificity	Induced in:	Humoral/ cell mediated	Specificity	
Melanoma	BALB/c mouse	Mono-clonal	Melanoma-asso-ciated proten (97 kDa)	Rabbit	Poly-clonal	Combining site-associated CRI on anti-p97 antibod-ies with same epitope speci-ficity as Ab_1	BALB/c mouse	Ab_3 (polyclo-nal) DTH	p97 antigen Melanoma cells	76
Melanoma	BALB/c mouse	Mono-clonal	HMW-MAA	BALB/c mouse	Mono-clonal	Combining site-associated idiotype on homologous Ab_1	BALB/c mouse Human	Ab_3 (polyclo-nal) Ab_3 (polyclo-nal)	HMW-MAA Unknown	55, 56
Gastric car-cinoma	BALB/c mouse	Mono-clonal	Glycoprotein on gastrointestinal carcinoma (30–40 kDa)	Goat	Poly-clonal	Combining site-associated idiotype on homologous Ab_1	BALB/c mouse Rabbit	Ab_3 (mono-clonal) Ab_3 (polyclo-nal)	Glycoprotein on gastro-intestinal carcinoma (30–40 kDa)	37
Colon carci-noma	BALB/c mouse	Mono-clonal	Glycoprotein on gastrointestinal carcinoma (30–40 kDa)	Goat	Poly-clonal	Combining site-associated idiotype on homologous Ab_1	Rabbit Human	Ab_3 (polyclo-nal) Ab_3 (polyclo-nal)	Glycoprotein on gastro-intestinal carcinoma (30–40 kDa)	36, 41

T-cell leukemia	BALB/c mouse	Mono-clonal	Glycoprotein on T-cell leukemia (37 kDa)	BALB/c mouse	Mono-clonal	Combining site-associated idiotype homologous Ab$_1$ and heterologous anti-gp37 antibody	BALB/c mouse Rabbit	Ab$_3$ (polyclonal) Ab$_3$ (polyclonal)	Glycoprotein on T-cell leukemia (37 kDa)	3, 4
Mammary carcinoma	BALB/c mouse	Mono-clonal	Globoside on mammary carcinoma (CAMBr1)	BALB/c mouse	Mono-clonal	Combining site-associated idiotype on homologous Ab$_1$	BALB/c mouse Rabbit	Ab$_3$ (monoclonal) Ab$_3$ (polyclonal)	Globoside on mammary carcinoma (CAMBr1)	112
Breast carcinoma	Human	Mono-clonal	Glycoprotein on breast carcinoma (gp60–68)	Rabbit	Poly-clonal	Combining site-associated idiotype on homologous Ab$_1$	C3He mouse	Ab$_3$ (polyclonal) DTH	MMTV expressing cross-reactive (with gp60–68) antigen Antigen-positive mammary carcinoma cells	103

^{a} Abbreviations: CRI, cross-reactive idiotope; HMW-MAA, high-molecular-weight melanoma-associated antigen; MMTV, mouse mammary tumor virus.

Both Ab_1 detect the same 30- to 40-kilodalton (kDa) glycoprotein on tumor cells (27, 90). MAb-binding competition studies indicated that although both Ab_1 bind the same antigen, they most probably recognize different epitopes (90). The two Ab_2 preparations, produced in goats against each of the two Ab_1, were extensively studied in vitro (36, 37). They completely inhibited the binding of Ab_1 to target cells. A major population of the polyclonal Ab_2 bound within or close to the combining site of the Ab_1. In vivo, the Ab_2 induced monoclonal Ab_3 in BALB/c mice and polyclonal Ab_3 in rabbits. In both species, the Ab_3 bound to tumor antigen isolated on Ab_1 immunoaffinity columns and showed the same binding specificities to various tumor cells as the corresponding Ab_1 did. Ab_3 elicited by Ab_2 to Ab_1 CO17-1A and GA733 lysed colon carcinoma cells in culture in conjunction with complement or macrophages, respectively. The Ab_2 against Ab_1 CO17-1A already have been used in clinical trials (see Clinical Trials with Ab_2). Recently, we have also developed a monoclonal rat Ab_2 against Ab_1 CO17-1A. The Ab_2 induced antigen-binding Ab_3 across species barriers. Ab_3 were measurable in unprocessed sera (D. Herlyn, M. Prewett, and M. Sperlag, unpublished results). In contrast, after immunization of rabbits with the polyclonal goat Ab_2 described above, the Ab_3 had to be purified from rabbit sera to demonstrate their specific binding to isolated tumor antigen (36, 37). These studies indicate the superiority of monoclonal Ab_2 over polyclonal Ab_2, both raised against the same Ab_1, in their capacity to induce antigen-specific Ab_3 across species barriers.

Bhattacharya-Chatterjee et al. have generated a syngeneic monoclonal cascade of $Ab_1 \rightarrow Ab_2 \rightarrow Ab_3$. The Ab_1 is directed against the human T-cell leukemia antigen gp37 (3, 4). This antigen is present at high density on the surfaces of human T-cell leukemia cells and has been partially purified. Syngeneic monoclonal Ab_2 was produced in BALB/c mice. It inhibited the binding of Ab_1 to either semipurified gp37 or antigen-positive target cells. Moreover, the binding of Ab_1 to Ab_2 was inhibited by antigen, and the Ab_2 reacted not only with the homologous Ab_1 but also with xenogeneic rabbit antibodies specific for the same antigen. In vivo, Ab_2 induced antigen-binding Ab_3 in mice and rabbits in the absence of antigen.

Another syngeneic cascade ($Ab_1 \rightarrow Ab_2 \rightarrow Ab_3$) was described by Kusama et al. in the human melanoma system (55, 56). The Ab_1 binds to a high-molecular-weight melanoma-associated antigen. Syngeneic Ab_2 specifically inhibits the binding of Ab_1 to melanoma cells and induces Ab_3 that inhibits the binding of Ab_2 to Ab_1, competes with Ab_1 for binding to melanoma cells, and immunoprecipitates the same antigen as the Ab_1. However, it is unclear whether the Ab_2 bears an internal image of the antigen, since immunizations across species barriers have not been performed. Nevertheless, the Ab_2 were used in clinical trials (see Clinical Trials with Ab_2).

Viale et al. (112) have studied a saccharide epitope (CAMBr1) of human tissue-specific tumor-associated globoside that is present on a human mammary carcinoma cell line and is associated with poor prognosis of breast cancer. A monoclonal Ab_1 to CAMBr1 was used to produce two syngeneic Ab_2 in mice. Functional and biological homologies of the Ab_2 with antigen were demonstrated both in vitro and in vivo by using Ab_2 as the immunogen across species barriers. Rabbits immunized with the Ab_2 produced antigen-binding Ab_3 that showed the same pattern of reactivity to tumor cells as the Ab_1 did.

Smorodinsky et al. reported an anti-idiotypic cascade in a human mammary carcinoma system (103). Two Ab_1 are directed against the 60- and 68-kDa antigens expressed in human mammary carcinoma. These antigens cross-react with the major mouse mammary tumor virus antigens gp52 and gp36. Rabbit Ab_2 were produced against each of the two Ab_1. The Ab_2 mimicked in vitro the gp60 and gp68 antigens and induced specific cellular and humoral responses in mice. Ab_2 immunizations across species barriers have not been performed in those studies. Because the antigen is also expressed in murine mammary tumors, this particular model offers the possibility of studying the protective effects of Ab_2 immunizations against tumor challenge in mice.

It has been demonstrated that Ab_2 immunizations can induce specific humoral responses to human tumors across species barriers. However, in most systems, cytotoxic reactivities of Ab_3 responses have not been demonstrated. Therefore, the beneficial role of these responses remains unclear. Furthermore, except for human tumor antigens that cross-react with mouse tumor antigens (103), it is not possible, in any of the other systems studied, to investigate cellular and protective immune responses induced by Ab_2 immunizations.

Clinical Trials with Ab_2

In a phase I clinical trial (41), 30 patients with advanced colorectal carcinoma were treated with alum-precipitated (0.5 to 4 mg per injection) goat Ab_2, which had been shown in preclinical studies to functionally mimic a gastrointestinal carcinoma-associated antigen (36, 37) (see Experimental Induction of Immunity to Human Tumors by Ab_2 Immunizations). Most of the 30 patients developed local erythema at the site of intradermal or subcutaneous injection of Ab_2. However, none of the patients experienced systemic adverse effects. All patients developed Ab_3 with Ab_1-like binding specificities on the surfaces of cultured tumor cells. Ab_3 also competed with Ab_1 for binding to colorectal carcinoma cells. After elution of the serum immunoglobulins from colorectal carcinoma cells, fractions of Ab_3-containing sera bound to purified tumor antigen and inhibited the binding of

Ab_2 to Ab_1. Therefore, Ab_3 bind to the same isolated antigen as Ab_1 do, and may share idiotopes with Ab_1. Of the 30 treated patients, 6 showed partial remission (>20% decrease in tumor metastasis for more than 4 weeks), 7 maintained a stable disease (no change in tumor size for more than 4 weeks), and 17 showed increases in tumor masses. However, 9 of the 13 patients who improved clinically after Ab_2 therapy had also received chemotherapy. A beneficial role of Ab_2 in the four patients treated with Ab_2 alone is suggested. In future studies we will investigate the cellular immune responses of cancer patients treated with polyclonal goat Ab_2.

Ab_2 also have been administered to melanoma patients in an active immunotherapy trial (47). The preclinical studies with the Ab_2 have been described in detail above (see Experimental Induction of Immunity to Human Tumors by Ab_2 Immunization) and have been published elsewhere (55, 56). Although the Ab_2 elicited in the patients Ab_3 that bound specifically to Ab_2, the Ab_3 did not bind the melanoma-associated antigen defined by the Ab_1.

PASSIVE IMMUNOTHERAPY OF CANCER WITH Ab_2

In the particular case of lymphoma, Ab_2 can be used for passive immunotherapy since the TAA is an immunoglobulin (idiotope). The beneficial role of Ab_2 in the treatment of lymphoma was described in 1972 by Lynch et al. (62) for animal models. They showed that administration of Ab_2 directed against myeloma idiotype proteins suppresses the growth of the tumors in mice.

Nadler et al. (71) pioneered the treatment of human B-cell lymphoma with Ab_2. They treated one patient but did not obtain any clinical response. Miller and co-workers were more successful (60, 68, 69) in their study, which initially included 13 patients. One patient has been in complete remission for more than 6 years after Ab_2 treatment. Of the other 12 patients, 6 had a clinically significant partial remission of their disease.

Although these studies look promising, several obstacles are associated with the low rate of success in immunotherapy of B-cell lymphoma with Ab_2. The major obstacles are (i) the presence of circulating antigen (idiotype-positive immunoglobulin); (ii) the appearance of variant tumor clones that express an idiotype different from that of the primary tumor (antigenic modulation); and (iii) the immune responses of the patients against the Ab_2.

Gordon et al. (26) and Nadler et al. (71) reported shedding of idiotype-positive immunoglobulin into the circulation of patients with B-cell lymphoma. Circulating idiotype can decrease the efficacy of the Ab_2 treatment by blocking Ab_2 binding to idiotype-bearing tumor cells. Furthermore, immune complexes between Ab_2 and idiotype-positive immunoglobulin may exert toxic side effects. Antigenic modulation of the lymphoma-associated idiotype in response to Ab_2 therapy has led to the emergence of variant tumor cells which have lost idiotype expression and therefore have resisted any further treatment with Ab_2 (65, 82).

The most common mechanism underlying the emergence of variant tumor cells during Ab_2 treatment is probably somatic mutation within the idiotype-encoding region of the V gene to generate a new idiotype (66). Thus, the use of a series of Ab_2 with various specificities will have to be considered for treatment of lymphomas.

Induction of an immune response (to the xenogeneic Ab_2) in the patients has been described as a major obstacle to the immunotherapy of cancer with MAbs (Ab_1 or Ab_2 [68]). These responses may inactivate the injected MAbs. The administration of large initial doses of MAb may suppress this response (97). To minimize human immune responses against the administered MAb, chimeric antibodies with the human Fc fragment and the murine specific variable region have been used in clinical trials (52).

Despite the difficulties encountered in immunotherapy of lymphomas with Ab_2, encouraging results have been obtained in treated patients. Ab_2 therapy could be useful especially when the tumor is resistant to conventional therapy.

CONCLUSIONS

The studies in experimental tumor systems reviewed here clearly show that Ab_2 immunizations can induce specific humoral, cellular, and protective antitumor immunity. The demonstration of antigen-specific humoral immunity induced by Ab_2 in cancer patients (41) has allowed new approaches to immunotherapy of cancer. It is possible that priming of B cells by patient tumor cells facilitated the induction of antigen-specific Ab_3 by Ab_2 immunizations. Although analysis of the immune responses of these patients yielded encouraging results, it is too early to draw definitive conclusions about the role of these immune responses in the control of the disease.

Each of the Ab_2 used to generate antitumor responses in the various systems described is directed to B-cell idiotopes. Thus, Ab_2 to B cell idiotopes induced both humoral and cellular immunity. However, the studies of the Sendai virus (17) and *Listeria* (48) systems suggest that in outbred human populations, clonotypic antibodies to the antigen receptor on human T-cell clones are prime candidates for elicitation of tumoricidal T-cell responses. The feasibility of this approach has been demonstrated in various antigenic systems in which soluble antigen or autologous antigen-positive cells stimulated antigen-specific human T-cell clones (86, 110, 114). Furthermore, clonotypic antibodies were induced against antigen-specific human T-cell clones (66).

Many important questions remain unanswered in the studies reviewed here. Systematic evaluation of the effects of a number of variables on the outcome of Ab_2 immunizations by using a single well-defined antigenic system is essential for future consideration of clinical applications of Ab_2. Important variables include (i) combining-site- versus non-combining-site-specific Ab_2; (ii) monoclonal versus polyclonal Ab_2; (iii) the type of adjuvant used; and (iv) the form of Ab_2. From the studies described in detail above (see Selection of Ab_2), it seems that combining-site-related monoclonal Ab_2 ($Ab_2\beta$, internal-image Ab_2) are prime candidates for clinical trials. Aluminum hydroxide is the choice of adjuvant at present, in terms of both specificity and potential for human use (41). Various forms of Ab_2 [immunoglobulin G, $F(ab')_2$ fragments, $F(ab')_2$ fragments coupled to immunogenic carriers; see Ab_2 Administration Schedule] must be compared in experimental immunizations of animals to induce maximal immunity.

In comparative experimental immunizations with Ab_2 and antigen, Ab_2 have been either equal to (5, 83) or less effective than (17, 80, 106) antigen in their capacity to induce antigen-specific immune responses. The numerous reports on the lower efficiency of Ab_2 are not surprising, since both the selection of Ab_2 and the immunization schedules must be optimized. Furthermore, monoclonal Ab_2 usually mimic only one epitope (since they usually are directed against monoclonal Ab_1), whereas antigen contains several epitopes. Therefore, a mixture of several monoclonal Ab_2 mimicking various epitopes has the potential of being as effective as antigen in inducing antigen-specific immune responses. The immunogenicities of Ab_2 and antigen will ultimately have to be compared in patients who usually are tolerant to the TAA expressed on their growing tumors (30, 113). Successful cloning of genes encoding various TAAs (9, 28, 43, 46, 61, 89) provides the basis for future comparative clinical trials with Ab_2 and recombinant antigen.

Acknowledgments. We thank Shirley Peterson for excellent editorial assistance and Andrea Barol for typing this manuscript.

These studies are supported by Public Health Service grant CA-43735 from the National Institutes of Health.

REFERENCES

1. **Abdou, N. I., H. Wall, H. B. Lindsley, J. F. Hasley, and T. Suzuki.** 1981. Network theory in autoimmunity. *In vitro* suppression of serum anti-DNA antibody binding to DNA by anti-idiotypic antibody to lupus erythematosus. *J. Clin. Invest.* **67:**1297–1304.
2. **Antel, J., J. J. Oger, S. Jackevicius, H. H. Kuo, and G. W. Arnason.** 1982. Modulation of

T-lymphocyte differentiation antigens: potential relevance for multiple sclerosis. *Proc. Natl. Acad. Sci. USA* **79**:3330–3334.

3. **Bhattacharya-Chatterjee, M., S. K. Chatterjee, S. Vasile, B. K. Seon, and H. Kohler.** 1988. Idiotype vaccines against human T cell leukemia. II. Generation and characterization of a monoclonal idiotype cascade (Ab1, Ab2, and Ab3). *J. Immunol.* **141**:1398–1403.

4. **Bhattacharya-Chatterjee, M., M. W. Pride, B. K. Seon, and H. Kohler.** 1987. Idiotype vaccines against human T cell acute lymphoblastic leukemia. *J. Immunol.* **139**:1354–1360.

5. **Bhogal, B. S., K. H. Nollstadt, Y. D. Karkhanis, D. M. Schmatz, and E. B. Jacobsen.** 1988. Anti-idiotypic antibody with potential use as an *Eimeria tenella* sporozoite antigen surrogate for vaccination of chickens against coccidiosis. *Infect. Immun.* **56**:1113–1119.

6. **Bona, C. A.** 1984. Parallel sets and the internal image of antigen within the idiotypic network. *Fed. Proc.* **43**:2558–2562.

7. **Bosslet, K., M. Buchler, R. Klapdor, C. Muhrer, H. H. Sedlacek, and G. Schulz.** 1988. Human monoclonal anti-idiotypic antibodies as an epitope vaccine against pancreatic carcinoma. *Behring Inst. Mitt.* **82**:193–196.

8. **Bruck, C., M. S. Co, M. Slaoui, G. N. Gaulton, T. Smith, B. N. Fields, J. I. Mullins, and M. I. Greene.** 1986. Nucleic acid sequence of an internal image-bearing monoclonal anti-idiotype and its comparison to the sequence of the external antigen. *Proc. Natl. Acad. Sci. USA* **83**:6578–6582.

9. **Chao, M. V., M. A. Bothwell, A. H. Ross, H. Koprowski, A. A. Lanahan, C. R. Buck, and A. Sehgal.** 1986. Gene transfer and molecular cloning of the human NGF receptor. *Science* **232**:518–521.

10. **Chatenoud, L., and J. F. Bach.** 1984. Antigenic modulation—a major mechanism of antibody action. *Immunol. Today* **5**:20–25.

11. **Courtenay-Luck, N. S., A. A. Epenetos, G. B. Sivolapenko, M. Larche, J. R. Barkans, and M. A. Ritter.** 1988. Development of anti-idiotypic antibodies against tumor antigens and autoantigens in ovarian cancer patients treated intraperitoneally with mouse monoclonal antibodies. *Lancet* **ii**:894–896.

12. **Dekruyff, R. H., S.-T. Ju, A. J. Hunt, T. R. Mosmann, and D. T. Umetsu.** 1989. Induction of antigen-specific antibody responses in primed and unprimed B cells. Functional heterogeneity and T_H1 and T_H2 T cell clones. *J. Immunol.* **142**:2575–2582.

13. **DeSaint Basile, G., A. Durandy, G. Somme, and C. Griscelli.** 1987. Idiotypy of human anti-candida albicans antibodies: recurrence, presence of a crossreactive autoanti-idiotypic-like activity, and role of the induction of specific *in vitro* antibody response. *J. Immunol.* **138**:417–422.

14. **Dunn, P. L., C. A. Johnson, J. M. Styles, S. S. Pease, and C. J. Dean.** 1987. Vaccination with syngeneic monoclonal anti-idiotype protects against a tumor challenge. *Immunology* **60**:181–186.

15. **Ertl, H. C., and C. Bona.** 1988. Criteria to define anti-idiotypic antibodies carrying the internal image of an antigen. *Vaccine* **6**:80–84.

16. **Ertl, H. C., J. Homans, S. Tournas, and R. W. Finberg.** 1984. Sendai virus-specific T cell clones. V. Induction of a virus-specific response by anti-idiotypic antibodies directed against a helper T cell clone. *J. Exp. Med.* **159**:1778–1783.

17. **Ertl, H. C. J., and R. W. Finberg.** 1984. Sendai virus-specific T cell clones: induction of cytolytic T cells by an anti-idiotypic antibody directed against a helper T cell clone. *Proc. Natl. Acad. Sci. USA* **81**:2850–2854.

18. **Foley, E. J.** 1953. Antigenic properties of methylcholenthrene-induced tumors in mice of the strain of origin. *Cancer Res.* **13**:835–839.

19. **Forstrom, J. W., K. A. Nelson, G. T. Nepom, I. Hellstrom, and K. E. Hellstrom.** 1983. Immunization to a syngeneic sarcoma by a monoclonal auto-anti-idiotypic antibody. *Nature* (London) **303**:627–629.

20. **Francotte, M., and J. Urbain.** 1984. Induction of anti-tobacco mosaic virus antibodies in mice by rabbit anti-idiotypic antibodies. *J. Exp. Med.* **160**:1485–1494.

21. **Francotte, M., and J. Urbain.** 1985. Enhancement of antibody response by mouse dendritic cells pulsed with tobacco mosaic virus or with rabbit antiidiotypic antibodies raised against a private rabbit idiotype. *Proc. Natl. Acad. Sci. USA* **82**:8149–8152.

22. **Gaulton, G. N., A. H. Sharpe, D. W. Chang, B. N. Fields, and M. I. Greene.** 1986. Syngeneic monoclonal internal image anti-idiotypes as prophylactic vaccines. *J. Immunol.* **137**:1743–1749.

23. **Geha, R. S.** 1982. Presence of auto-anti-idiotypic antibody during the normal human immune response to tetanus toxoid antigen. *J. Immunol.* **129**:139–144.

24. **Geltner, D., E. Franklin, and B. Frangione.** 1980. Antiidiotypic activity in the IgM fractions of mixed cryoglobulins. *J. Immunol.* **125**:1530–1535.

25. **Gonzalez-Cabrero, J., J. Egido, J. Sancho, and F. Moldenhauer.** 1986. Presence of shared idiotypes in serum and immune complexes in patients with IgA nephropathy. *Clin. Exp. Immunol.* **68**:694–702.

26. **Gordon, J., A. K. Abdul-Ahad, T. J. Hamblin, F. K. Stevenson, and G. T. Stevenson.** 1984. Mechanisms of tumour cell escape encountered in treating lymphocytic leukaemia with anti-idiotypic antibody. *Br. J. Cancer* **49**:547–557.

27. **Gottlinger, H. G., I. Funke, J. P. Johnson, J. M. Gokel, and G. Riethmuller.** 1986. The epithelial cell surface antigen 17-1A, a target for antibody mediated tumor therapy: its biochemical nature, tissue distribution and recognition by different monoclonal antibodies. *Int. J. Cancer* **38**:47–53.

28. **Graf, L. H., C. D. Rosenberg, V. Mancino, and S. Ferrone.** 1988. Transfer and coamplification of a gene encoding a 96-kda immune IFN-inducible human melanoma-associated antigen. *J. Immunol.* **141**:1054–1060.

29. **Greene, M., M. Nelles, M. Sy, and A. Nisonoff.** 1982. Regulation of immunity to azobenzenearsonate hapten. *Adv. Immunol.* **32**:253–257.

30. **Hamby, C. V., S.-K. Liao, T. Kanamaru, and S. Ferrone.** 1987. Immunogenicity of human melanoma-associated antigens defined by murine monoclonal antibodies in allogeneic and xenogeneic hosts. *Cancer Res.* **47**:5284–5289.

31. **Hellstrom, K. E., I. Hellstrom, and G. T. Nepom.** 1977. Specific blocking factors—are they important? *Biochim. Biophys. Acta* **473**:121–148.

32. **Herlyn, D., M. Herlyn, Z. Steplewski, and H. Koprowski.** 1985. Monoclonal antihuman tumor antibodies of six isotypes in cytotoxic reactions with human and murine effector cells. *Cell. Immunol.* **92**:105–114.

33. **Herlyn, D., and H. Koprowski.** 1988. Anti-idiotypic antibodies in cancer therapy, p. 123–134. *In* C. A. Bona (ed.), *Biological Applications of Anti-Idiotypes.* CRC Press, Inc., Boca Raton, Fla.

34. **Herlyn, D., M. Lubeck, Z. Steplewski, and H. Koprowski.** 1985. Destruction of human tumors by IgG2a monoclonal antibodies and macrophages, p. 165–172. *In* R. A. Reisfeld and S. Sell (ed.), *Monoclonal Antibodies and Cancer Therapy.* Alan R. Liss, Inc., New York.

35. **Herlyn, D., J. Powe, A. H. Ross, M. Herlyn, and H. Koprowski.** 1985. Inhibition of human tumor growth by IgG2a monoclonal antibodies correlates with antibody density of tumor cells. *J. Immunol.* **134**:1300–1304.

36. **Herlyn, D., A. H. Ross, D. Iliopoulos, and H. Koprowski.** 1987. Induction of specific immunity to human colon carcinoma by anti-idiotypic antibodies to monoclonal antibody CO17-1A. *Eur. J. Immunol.* **17**:1649–1652.

37. **Herlyn, D., A. H. Ross, and H. Koprowski.** 1986. Anti-idiotypic antibodies bear the internal image of a human tumor antigen. *Science* **232**:100–102.

38. **Herlyn, D., Z. Steplewski, M. Herlyn, and H. Koprowski.** 1980. Inhibition of colorectal carcinoma tumor growth in nude mice by monoclonal antibody. *Cancer Res.* **40**:717–721.

39. **Herlyn, D., M. Wettendorff, D. Iliopoulos, and H. Koprowski.** 1988. Functional mimicry of tumor-associated antigens by anti-idiotypic antibodies. *Exp. Clin. Immunogenet.* **5**:165–175.

40. **Herlyn, D., M. Wettendorff, and H. Koprowski.** 1989. Modulation of cancer patients' immune responses by anti-idiotypic antibodies. *Int. Rev. Immunol.* **4**:347–357.

41. **Herlyn, D., M. Wettendorff, E. Schmoll, D. Iliopoulos, I. Schedel, U. Dreikhausen, R. Raab, A. H. Ross, H. Jaksche, M. Scriba, and H. Koprowski.** 1987. Anti-idiotype immunization of cancer patients: modulation of the immune response. *Proc. Natl. Acad. Sci. USA* **84**:8055–8059.

42. **Herlyn, M., Z. Steplewski, D. Herlyn, and H. Koprowski.** 1979. Colorectal carcinoma-specific antigen: detection by means of monoclonal antibodies. *Proc. Natl. Acad. Sci. USA* **76**:1438–1482.

43. **Hotta, H., A. H. Ross, K. Huebner, M. Isobe, S. Wendeborn, M. V. Chao, R. B. Ricciardi, Y. Tsujimoto, C. M. Croce, and H. Koprowski.** 1988. Molecular cloning and characterization of an antigen associated with early stages of melanoma tumor progression. *Cancer Res.* **48**:2955–2962.

44. **Infante, A. J., P. D. Infante, S. Gillis, and C. G. Fathman.** 1982. Definition of T cell idiotypes using anti-idiotypic antisera produced by immunization with T cell clones. *J. Exp. Med.* **155**:1100–1107.

45. **Jerne, N. K., J. Roland, and P. A. Cazenave.** 1982. Recurrent idiotypes and internal images. *EMBO J.* **1**:243–247.

46. **Johnson, D., A. Lanahan, C. R. Buck, A. Sehgal, C. Morgan, E. Mercer, M. Bothwell, and M. Chao.** 1986. Expression and structure of the human NGF receptor. *Cell* **47**:545–554.

47. **Kageshita, T., Z. J. Chen, J.-W. Kim, M. Kusama, U. M. Kekish, T. Trujillo, M. Temponi, A. Mittleman, and S. Ferrone.** 1988. Murine anti-idiotypic monoclonal antibodies to syngeneic antihuman high-molecular-weight melanoma-associated antigen monoclonal antibodies: development, characterization, and clinical applications. *Pigment Cell Res.* **1**(Suppl.):185–191.

48. **Kaufmann, S. H. E., K. Eichmann, J. Muller, and L. J. Wrazel.** 1985. Vaccination against the intracellular bacterium *Listeria monocytogenes* with a clonotypic antiserum. *J. Immunol.* **134**:4123–4127.

49. **Kaye, J. S., J. Porcelli, J. Tite, B. Jones, and C. A. Janeway.** 1983. Both a monoclonal antibody and antisera specific for determinants unique to individual cloned helper T cell lines can substitute for antigen and antigen presenting cells in the activation of T cells. *J. Exp. Med.* **158**:836–856.

50. **Kelsoe, G., M. Reth, and K. Rajewski.** 1981. Control of idiotope expression by monoclonal anti-idiotope and idiotope-bearing antibody. *Eur. J. Immunol.* **11**:418–423.

51. **Kennedy, R. C., G. R. Dreesman, J. S. Butel, and R. E. Lanford.** 1985. Suppression of *in vivo* tumor formation induced by simian virus 40-transformed cells in mice receiving antiidiotypic antibodies. *J. Exp. Med.* **161**:1432–1449.

52. **Khazaeli, M. B., K. Rogers, and A. T. LoBuglio.** 1989. Quantitations of chimeric mouse/human monoclonal antibody (CO17-1A) and human anti-mouse antibody response in serum of patients. *Proceedings of the 7th International Congress on Immunology.* Abstract no. 106-46, p. 754.

53. **Koprowski, H., and D. Herlyn.** 1986. Anti-idiotype vaccines against viral infections, p. 401–411. *In* A. L. Notkins and M. B. A. Oldstone (ed.), *Concepts in Viral Pathogenesis*, vol. 2. CRC Press, Inc., Boca Raton, Fla.

54. **Koprowski, H., D. Herlyn, M. Lubeck, E. DeFreitas, and H. F. Sears.** 1984. Human anti-idiotype antibodies in cancer patients: is the modulation of the immune response beneficial for the patient? *Proc. Natl. Acad. Sci. USA* **81**:216–219.

55. **Kusama, M., T. Kageshita, K. T. Kim, and S. Ferrone.** 1987. Antiidiotypic monoclonal antibodies to antihuman high-molecular-weight melanoma-associated antigen monoclonal antibodies. *Proc. Am. Assoc. Cancer Res.* **28**:361.

56. **Kusama, M., T. Kageshita, M. Tsujisaki, F. Perosa, and S. Ferrone.** 1987. Syngeneic antiidiotypic antisera to murine antihuman high-molecular-weight melanoma-associated antigen monoclonal antibodies. *Cancer Res.* **47**:4312–4317.

57. **Lee, V. K., T. M. Harriot, V. K. Kuchroo, W. J. Halliday, I. Hellstrom, and K. E. Hellstrom.** 1985. Monoclonal antiidiotypic antibodies related to a murine oncofetal bladder tumor antigen induce specific cell-mediated tumor immunity. *Proc. Natl. Acad. Sci. USA* **82**:6286–6290.

58. **Lee, V. K., K. E. Hellstrom, and G. T. Nepom.** 1986. Idiotypic interactions in immune responses to tumor-associated antigens. *Biochim. Biophys. Acta* **865**:127–139.

59. **Lefvert, A. K.** 1987. Idiotypes and anti-idiotypes of human autoantibodies to the acetylcholine receptor in myasthenia gravis. *Monogr. Allergy* **22**:57–70.

60. **Levy, R., and R. A. Miller.** 1983. Tumor therapy with monoclonal antibodies. *Fed. Proc.* **42**:2653–2656.

61. **Linnenbach, A. J., J. Wojcierowski, S. Wu, J. J. Pyrc, A. H. Ross, B. Dietzschold, D. Speicher, and H. Koprowski.** 1989. Sequence investigation of the major gastrointestinal tumor-associated antigen gene family, GA733. *Proc. Natl. Acad. Sci. USA* **86**:27–31.

62. **Lynch, R. C., R. J. Craff, S. Sirishina, E. S. Simms, and H. N. Eisen.** 1972. Myeloma proteins as tumor-specific transplantation antigens. *Proc. Natl. Acad. Sci. USA* **69**:1540–1545.

63. **Margalit, H., J. L. Spouge, J. L. Cornette, K. B. Cease, C. Delisi, and J. A. Berzofsky.** 1987. Prediction of immunodominant helper T cell antigenic sites from the primary sequence. *J. Immunol.* **138**:2213–2229.

64. **Masouderis, S. P., M. J. Branks, and E. J. Victoria.** 1987. Antiidiotypic IgG crossreactive with Rh alloantibodies in red cell autoimmunity. *Blood* **70**:710–715.

65. **Meeker, T., J. Lowder, M. L. Cleary, S. Steward, R. Warnke, J. Sklar, and R. Levy.** 1985. Emergence of idiotype variants during treatment of B-cell lymphoma with anti-idiotype antibodies. *N. Engl. J. Med.* **312**:1658–1665.

66. **Meuer, S. C., K. A. Fitzgerald, R. E. Hussey, F. C. Hodgdon, S. F. Schlossman, and E. L. Rheinerz.** 1983. Clonotypic structures involved in antigen-specific human T cell function. Relationship to the T_3 molecular complex. *J. Exp. Med.* **157**:705–719.

67. **Miller, G. G. P., P. I. Nadler, R. J. Hodes, and D. H. Sachs.** 1982. Modification of T cell antinuclease idiotype expression by *in vivo* administration of anti-idiotype. *J. Exp. Med.* **155**:190–200.

68. **Miller, R. A., J. Lowder, T. C. Meeker, S. Brown, and R. Levy.** 1987. Anti-idiotypes in B-cell tumor therapy. *NCI Monogr.* **3**:131–134.

69. **Miller, R. A., D. G. Maloney, R. Warnke, and R. Levy.** 1982. Treatment of B-cell lymphoma with monoclonal anti-idiotype antibody. *N. Engl. J. Med.* **306**:517–522.

70. **Mosier, D. E., R. T. Gulizia, S. M. Baird, and D. B. Wilson.** 1988. Transfer of a functional human immune system to mice with severe combined immunodeficiency. *Nature* (London) **335**:256–259.

71. **Nadler, L. M., P. Stashenko, R. Hardy, W. D. Kaplan, L. N. Button, D. W. Kufe, K. H. Antman, and S. F. Schlossman.** 1980. Serotherapy of a patient with a monoclonal antibody directed against a human lymphoma-associated antigen. *Cancer Res.* **40**:3147–3154.

72. **Nelson, K., and G. T. Nepom.** 1986. Idiotypic networks in tumor immunity, p. 177–185.

In G. Hoffman, J. Levy, and G. Nepom (ed.), *Paradoxes in Immunology.* CRC Press, Inc., Boca Raton, Fla.

73. **Nelson, K. A., E. George, C. Swenson, J. W. Forstrom, and K. E. Hellstrom.** 1987. Immunotherapy of murine sarcomas with auto-antiidiotypic monoclonal antibodies which bind to tumor-specific T cells. *J. Immunol.* **139**:2110–2117.

74. **Nepom, G. T., I. Hellstrom, and K. E. Hellstrom.** 1983. Suppressor mechanisms in tumor immunity. *Experientia* **39**:235–242.

75. **Nepom, G. T., and K. E. Hellstrom.** 1987. Anti-idiotypic antibodies and the induction of specific tumor immunity. *Cancer Metastases Rev.* **6**:489–502.

76. **Nepom, G. T., K. A. Nelson, S. L. Holbeck, I. Hellstrom, and K. E. Hellstrom.** 1984. Induction of immunity to a human tumor marker *in vivo* by administration of anti-idiotypic antibodies in mice. *Proc. Natl. Acad. Sci. USA* **81**:2864–2867.

77. **Nisonoff, A., and E. Lamoy.** 1981. Implications of the presence of an internal image of the antigen in anti-idiotypic antibodies: possible application to vaccine production. *Clin. Immunol. Immunopathol.* **21**:397–406.

78. **Old, L. J.** 1981. Cancer immunology: the search for specificity. G. H. A. Clowes Memorial Lecture. *Cancer Res.* **41**:361–375.

79. **Ollier, P., J. Rocca-Serra, G. Somme, J. Theze, and M. Fougereau.** 1985. The idiotypic network and the internal image: possible regulation of a germ-line network by paucigene encoded Ab2 (anti-idiotypic) antibodies in the GAT system. *EMBO J.* **4**:3681–3688.

80. **Phillips, S. M., P. J. Perrin, D. J. Walker, N. G. Fathelbab, G. P. Linette, and M. A. Idris.** 1988. The regulation of resistance to *Schistosoma mansoni* by auto-anti-idiotypic immunity. *J. Immunol.* **141**:1728–1733.

81. **Powell, T. J., Jr., R. Spann, M. Vakil, J. F. Kearney, and E. W. Lamon.** 1988. Activation of a functional idiotype network response by monoclonal antibody specific for a virus (M-MuLV)-induced tumor antigen. *J. Immunol.* **140**:3266–3272.

82. **Raffeld, M., L. Neckers, D. L. Longo, and J. Cossman.** 1985. Spontaneous alteration of idiotype in a monoclonal B-cell lymphoma. Escape from detection by anti-idiotype. *N. Engl. J. Med.* **312**:1653–1658.

83. **Raychaudhury, S., Y. Saeki, J.-J. Chen, H. Iribe, H. Fuji, and H. Kohler.** 1987. Tumor-specific idiotype vaccines. II. Analysis of the tumor-related response induced by the tumor and by internal image antigens (Ab2β). *J. Immunol.* **139**:271–278.

84. **Raychaudhury, S., Y. Saeki, J.-J. Chen, and H. Kohler.** 1987. Tumor-specific idiotype vaccines. III. Induction of T helper cells by anti-idiotype and tumor cells. *J. Immunol.* **139**:2096–2102.

85. **Raychaudhury, S., Y. Saeki, H. Fuji, and H. Kohler.** 1986. Tumor-specific idiotype vaccines. I. Generation and characterization of internal image tumor antigen. *J. Immunol.* **137**:1743–1749.

86. **Rees, A. D. M., K. Praputpittaya, A. Scoging, N. Dobson, J. Ivanyi, D. Young, and J. R. Lamb, Jr.** 1987. T cell activation by anti-idiotypic antibody: evidence for the internal image. *Immunology* **60**:389–393.

87. **Rees, A. D. M., A. Scoging, N. Dobson, K. Praputpittaya, D. Young, J. Ivanyi, and J. R. Lamb, Jr.** 1987. T cell activation by anti-idiotypic antibody: mechanism of interaction with antigen-reactive T cells. *Eur. J. Immunol.* **17**:197–201.

88. **Ritz, J., J. M. Pesando, J. Notis-McConarty, and S. F. Schlossman.** 1980. Modulation of human acute lymphoblastic leukemia antigen induced by monoclonal antibody *in vitro. J. Immunol.* **125**:1506–1514.

89. **Rose, T. M., G. D. Plowman, D. B. Teplow, W. J. Dreyer, K. E. Hellstrom, and J. P. Brown.** 1986. Primary structure of the human melanoma-associated antigen p97 (melanotransferrin) deduced from the mRNA sequence. *Proc. Natl. Acad. Sci. USA* **83**:1261–1265.

90. **Ross, A. H., D. Herlyn, D. Iliopoulos, and H. Koprowski.** 1986. Isolation and characterization of a carcinoma-associated antigen. *Biochem. Biophys. Res. Commun.* **135:**297–303.

91. **Sacks, D. H.** 1987. Immunization against parasitic protozoa using anti-idiotypic antibodies. *Monogr. Allergy* **22:**166–171.

92. **Saeki, Y., J.-J. Chen, L. Shi, and H. Kohler.** 1989. Idiotypic intramolecular help. Induction of tumor-specific antibodies by monoclonal anti-idiotypic antibody with the help of Fc-specific T helper clones. *J. Immunol.* **142:**2629–2634.

93. **Saeki, Y., J.-J. Chen, L. Shi, S. Raychaudhury, and H. Kohler.** 1989. Characterization of "regulatory" idiotope-specific T cell clones to a monoclonal anti-idiotypic antibody mimicking a tumor-associated antigen (TAA). *J. Immunol.* **142:**1046–1052.

94. **Saxon, A., and E. Barnett.** 1984. Human auto-antiidiotypes regulating T cell-mediated reactivity to tetanus toxoid. *J. Clin. Invest.* **73:**342–348.

95. **Schwartz, R.** 1985. T-lymphocyte recognition of antigen in association with gene products of the major histocompatibility complex. *Annu. Rev. Immunol.* **3:**237–261.

96. **Sears, H. F., B. Atkinson, J. Mattis, C. Ernst, D. Herlyn, Z. Steplewski, P. Hayry, and H. Koprowski.** 1982. The use of monoclonal antibody in a phase I clinical trial of human gastrointestinal tumors. *Lancet* **i:**762–765.

97. **Sears, H. F., D. J. Bagli, D. Herlyn, E. DeFreitas, H. Suzuki, G. Steele, and H. Koprowski.** 1987. Human immune response to monoclonal antibody administration is dose-dependent. *Arch. Surg.* **122:**1384–1388.

98. **Sege, K., and P. A. Peterson.** 1978. Use of anti-idiotypic antibodies as cell-surface receptor probes. *Proc. Natl. Acad. Sci. USA* **75:**2443–2447.

99. **Sharpe, A. H., G. N. Gaulton, K. K. McDade, B. N. Fields, and M. I. Greene.** 1984. Syngeneic monoclonal anti-idiotype can induce cellular immunity to reovirus. *J. Exp. Med.* **160:**1195–1205.

100. **Shoelson, S. E., S. Marshall, H. Horikoshi, O. G. Kolterman, and H. Rubinstein.** 1986. Antiinsulin receptor antibodies in an insulin-dependent diabetic may arise as autoantiidiotypes. *J. Clin. Endocrinol. Metab.* **63:**56–61.

101. **Sikorska, H. M.** 1988. Therapeutic applications of anti-idiotypic antibodies. *J. Biol. Response Modif.* **7:**327–358.

102. **Singhai, R., and J. G. Levy.** 1987. Isolation of a T cell clone that reacts with both antigen and anti-idiotype: evidence for anti-idiotype as internal image for antigen at the T cell level. *Proc. Natl. Acad. Sci. USA* **84:**3836–3840.

103. **Smorodinsky, N. I., Y. Ghendler, R. Bakimer, S. Chaichuk, I. Keydar, and Y. Shoenfeld.** 1988. Towards an idiotype vaccine against mammary tumors. Induction of an immune response to breast cancer-associated antigens by anti-idiotypic antibodies. *Eur. J. Immunol.* **18:**1713–1718.

104. **Stein, K., and T. Soederstroem.** 1984. Neonatal administration of idiotype or anti-idiotype primes for protection against *Escherichia coli* K13 infection in mice. *J. Exp. Med.* **160:**1001–1011.

105. **Steinitz, M., S. Tamir, J.-E. Frodin, A.-K. Lefvert, and H. Mellstedt.** 1988. Human monoclonal anti-idiotypic antibodies. I. Establishment of immortalized cell lines from a tumor patient treated with mouse monoclonal antibodies. *J. Immunol.* **141:**3516–3522.

106. **Tanaka, M., N. Sasaki, and A. Seto.** 1986. Induction of antibodies against Newcastle disease virus with syngeneic anti-idiotype antibodies in mice. *Microbiol. Immunol.* **30:**323–331.

107. **Teitelbaum, D., J. Rauch, B. D. Stollar, and R. S. Schwartz.** 1984. *In vivo* effects of antibodies against a high frequency idiotype of anti-DNA antibodies in MRL mice. *Br. J. Immunol.* **132:**1282–1285.

108. **Townsend, A. R. M., J. Rothbard, F. M. Gotch, G. Bahadur, D. Wraith, and A. J. McMichael.** 1986. The epitopes of influenza nucleoprotein recognized by cytotoxic T lymphocytes can be defined with short synthetic peptides. *Cell* **44**:959–968.

109. **Urbain, J., and C. Wuilmart.** 1981. Idiotypic regulation in immune networks. *Contemp. Top. Mol. Immunol.* **8**:113–148.

110. **Uytdehaag, F., I. Claassen, H. Bunschoten, H. Loggen, T. Ottenhoff, V. Teeuwsen, and A. Osterhaus.** 1987. Human anti-idiotypic T-lymphocyte clones are activated by autologous anti-rabies virus antibodies presented in association with HLA-DQ molecules. *J. Mol. Cell. Immunol.* **3**:145–155.

111. **VanCleave, V. H., C. W. Naeve, and D. W. Metzger.** 1988. Do antibodies recognize amino acid side chains of protein antigens independently of the carbon backbone? *J. Exp. Med.* **167**:1841–1848.

112. **Viale, G., F. Grassi, M. Pelagi, R. Alzani, S. Menard, S. Miotti, R. Buffa, A. Gini, and A. G. Siccardi.** 1987. Anti-human tumor antibodies induced in mice and rabbits by "internal image" anti-idiotypic monoclonal immunoglobulins. *J. Immunol.* **139**:4250–4255.

113. **Wettendorff, M., D. Iliopoulos, M. Tempero, D. Kay, E. DeFreitas, H. Koprowski, and D. Herlyn.** 1989. Idiotypic cascades in cancer patients treated with monoclonal antibody CO17-1A. *Proc. Natl. Acad. Sci. USA* **86**:3787–3791.

114. **Zarling, J. M., J. W. Eichberg, P. A. Moran, J. McClure, P. Sridhar, and S.-L. Hu.** 1987. Proliferative and cytotoxic T cells to AIDS virus glycoproteins in chimpanzees immunized with a recombinant vaccinia virus expressing AIDS virus envelope glycoproteins. *J. Immunol.* **139**:988–990.

INDEX